Springer Complexity

Springer Complexity is an interdisciplinary program publishing the best research and academic-level teaching on both fundamental and applied aspects of complex systems – cutting across all traditional disciplines of the natural and life sciences, engineering, economics, medicine, neuroscience, social and computer science.

Complex Systems are systems that comprise many interacting parts with the ability to generate a new quality of macroscopic collective behavior the manifestations of which are the spontaneous formation of distinctive temporal, spatial or functional structures. Models of such systems can be successfully mapped onto quite diverse "real-life" situations like the climate, the coherent emission of light from lasers, chemical reaction-diffusion systems, biological cellular networks, the dynamics of stock markets and of the internet, earthquake statistics and prediction, freeway traffic, the human brain, or the formation of opinions in social systems, to name just some of the popular applications.

Although their scope and methodologies overlap somewhat, one can distinguish the following main concepts and tools: self-organization, nonlinear dynamics, synergetics, turbulence, dynamical systems, catastrophes, instabilities, stochastic processes, chaos, graphs and networks, cellular automata, adaptive systems, genetic algorithms and computational intelligence.

The three major book publication platforms of the Springer Complexity program are the monograph series "Understanding Complex Systems" focusing on the various applications of complexity, the "Springer Series in Synergetics", which is devoted to the quantitative theoretical and methodological foundations, and the "SpringerBriefs in Complexity" which are concise and topical working reports, case-studies, surveys, essays and lecture notes of relevance to the field. In addition to the books in these two core series, the program also incorporates individual titles ranging from textbooks to major reference works.

Robin R. Vallacher • Peter T. Coleman
Andrzej Nowak • Lan Bui-Wrzosinska • Larry
Liebovitch • Katharina G. Kugler • Andrea Bartoli

Attracted to Conflict: Dynamic Foundations of Destructive Social Relations

Springer

Robin R. Vallacher
Department of Psychology
Florida Atlantic University
Boca Raton, Florida
USA

Peter T. Coleman
International Center for Cooperation and Conflict Resolution
Teachers College
Columbia University
New York
USA

Andrzej Nowak
Department of Psychology
University of Warsaw
Warsaw, Poland

Lan Bui-Wrzosinska
International Center for Complexity and Conflict
Warsaw School of Social Sciences and Humanities
Warsaw, Poland

Larry Liebovitch
Division of Mathematics and Natural Sciences
City University of New York
Queens College
New York
USA

Katharina G. Kugler
Economic and Organisational Psychology
Ludwig-Maximilians-Universitaet Muenchen
Munich, Germany

Andrea Bartoli
School for Conflict Analysis and Resolution
George Mason University
Arlington, Virginia
USA

ISBN 978-3-642-35279-9 ISBN 978-3-642-35280-5 (eBook)
DOI 10.1007/978-3-642-35280-5
Springer Heidelberg New York Dordrecht London

Library of Congress Control Number: 2013935946

Printed on acid-free paper

Springer is part of Springer Science+Business Media (www.springer.com)

Preface

Conflict is ubiquitous in social life, leaving no realm of human interaction untouched. The potential for conflict in social relations has not gone unnoticed in psychology and the social sciences, as evident in the many insightful perspectives that have been advanced regarding this central feature of human nature. Despite this long-standing preoccupation, however, conflict in all its manifestations has yet to be integrated with respect to a unified set of principles. The lack of an agreed-upon theoretical synthesis is hardly surprising. Conflict can characterize social relations as distinct as marriage, parliamentary democracy, and international negotiations. Conflict can be a one-off experience, lasting minutes or even seconds, or it can become a persistent state that unfolds over months, years, or even decades. Conflict can be a constructive experience, promoting innovation and a shared reality, or it can engage the most destructive aspects of human nature, promoting extreme forms of violence, disrupting social progress, and magnifying differences in people's beliefs and values. And conflict can set in motion mechanisms that provide for its resolution or it can represent a self-sustaining process that makes resolution virtually impossible.

The perspective developed in this book was motivated in part by the inability of traditional models of social relations to impose coherence on the multifaceted nature of conflict in human affairs. We approached this task from what may seem like an ironic assumption: that the diversity of conflict can be integrated with respect to a perspective on science that encompasses the far greater diversity of phenomena in the natural world. This perspective—*nonlinear dynamical systems theory*—has proven successful since the 1970s in establishing the invariant processes that underlie seemingly distinct topics in fields such as physics, chemistry, cosmology, and biology. In recent years, the metaphors, principles, and methods of nonlinear dynamical systems have been successfully adapted to the subject matter of human experience, from cognitive and social psychology to economics and political science. This book describes how the dynamical perspective in psychology has been extended to understanding human conflict at different levels of social reality, from dyadic tensions to interstate warfare, with special emphasis on conflicts that are destructive and seemingly impervious to resolution. Such *intractable conflicts* are relatively rare, but their toll in loss of life, property damage, resource depletion, and draining of social capital qualifies them as among the most consequential of all social problems. And they have proven to be the most difficult to understand, let alone anticipate or resolve.

Considerable progress has been made over the past decade in framing the key features of intractable conflict in terms of formal models informed by nonlinear dynamical systems. These models have been instantiated in computer simulations and tested against archival and empirical data. But this book is intended to be heuristic as well as synthetic, establishing a road map for future research agendas. So in addition to framing conflict in dynamical terms and presenting supportive research, we point to areas in which more scholarly work is needed and we outline the strategies by which these theoretical and empirical goals can be accomplished.

Because the potential for destructive and intractable conflict cuts across all levels of human experience, comprehensive yet nuanced understanding is best served by enlisting the involvement of theorists, researchers, and practitioners with correspondingly diverse areas of expertise. The authors of this book reflect this multidisciplinary approach. Our team includes an experimental social psychologist (Vallacher), a social psychologist with expertise in computer simulation of social processes (Nowak), three social-organizational psychologists specializing in conflict management and resolution (Coleman, Bui-Wrzosinska, and Kugler), a cultural anthropologist with firsthand experience in intrastate conflict resolution and peace processes (Bartoli), and a physicist with expertise in complexity and nonlinear dynamical systems (Liebovitch). This collaborative effort, initiated in 2006, has tackled a wide range of topics, including dyadic (e.g., marital) conflicts, intra-organizational disputes, school violence, civil war, interstate warfare, negotiation, peace building, and sustainability.

The scholarly output of our team to date would have been impossible without the valuable cooperation of colleagues and the organizational and financial assistance provided by several institutes and foundations. Much of the initial work in developing the dynamical framework, generating formal models, and collecting empirical data was funded by a grant from the James S. McDonnell Foundation. Generous funding for conferences and workshops has been provided by the *International Center for Cooperation and Conflict Resolution* (ICCCR) and the *Advanced Consortium on Cooperation, Conflict, and Complexity* (AC4) at Columbia University; the *Institute for Conflict Analysis and Resolution* (ICAR) at George Mason University; the *Community Foundation of Boulder*; the *Peace Studies Program* at Florida Atlantic University; and the *Berghof Foundation*. Finally, we wish to acknowledge several colleagues, postdoctoral students, and graduate students for their invaluable scholarly contributions to our research program: Pawal Haltof, Wojciech Borkowski, Naira Musallam, Christine Chung, Jay Michaels, Susan Sullivan, and Vincent Naudot. The success of our efforts to date, as represented in this book, is attributable in large measure to the collaborative spirit—and constructive conflict—inspired in our team by the aforementioned individuals and organizational entities.

Boca Raton, FL	Robin R. Vallacher
New York, NY	Peter T. Coleman
Warsaw, Poland	Andrzej Nowak
Flushing, NY	Lan Bui-Wrzosinska
Munich, Germany	Larry Liebovitch
Arlington, VI	Katharina G. Kugler
	Andrea Bartoli

Contents

Chapter 1
Overview: Conflict in Human Experience

This is a book about conflict. But it is also a book about essential features of human nature that are expressed in every type of human interaction. In an even broader sense, this is a book about the basic processes that link conflict to a vast array of phenomena in the physical world. These seem like incompatible agendas. Conflict is not the only way humans interact, after all, and the conflicts that define human interactions would seem to have little in common with things like weather patterns, landslides, or bacterial growth. But as we shall see, science in recent years has exposed a set of basic operating rules that connect processes of all kinds in physical and social reality. This synthetic view is more than an abstraction; to the contrary, breakthroughs in mathematics, empirical methodology, and computer simulations have enabled scientists to identify the ways in which common processes and properties are manifest in very different phenomena. Our aim is to describe this new perspective and shine its concepts, methods, and tools on the recurrent and all-important issue of conflict in interpersonal, intergroup, and international relations.

Framing conflict in terms of generic processes would make little sense if conflict was viewed solely as dysfunctional, an anomaly of human interaction signaling a breakdown in the way people normally connect to one another. To be sure, conflict can showcase the very worst in people's thoughts, feelings, and actions with respect to one another. But conflict is far more than a problematic and unwanted feature of human experience. Quite the contrary, conflict is not only a frequent feature of social life, it is essential to our survival and progress as a species. Conflict, whether between individuals, groups, or cultures, is necessary for the construction of shared realities, technological and intellectual innovation, and adaptation to novel events and uncertain circumstances. Conflict is inherent in virtually every aspect of human encounter, from sport to parliamentary democracy, from fashion in the arts to paradigmatic challenges in the sciences, and from economic activity to intimate relationships.

Conflict in any of these realms, however, can become problematic, relinquishing its adaptive functions in favor of decidedly maladaptive consequences. Sports rivalries can precipitate violence; opposing political parties can become gridlocked, eliminating the possibility of compromise in framing public policy; clashes in art

R.R. Vallacher et al., *Attracted to Conflict: Dynamic Foundations of Destructive Social Relations*, Peace Psychology Book Series, DOI 10.1007/978-3-642-35280-5_1,

and science can stifle rather than inspire the exchange of ideas and information; economic competition can degenerate into sabotage within a society or into war between societies; and conflicting perspectives in an intimate relationship can spiral into distrust, antagonism, and physical abuse.

Conflict *per se*, then, is neither an exception nor a problem. Conflict becomes a problem—potentially the most serious social problem humans face—when it loses its constructive features and becomes defined in terms of its destructive potential, particularly when the conflict becomes protracted over time with no obvious means of resolution. Fortunately, enduring destructive conflicts—termed *intractable conflicts* (cf. Coleman, 2003)—are relatively rare, constituting about 5–7 % of all conflicts between individuals, groups, and countries (Coleman, 2011). But even here, conflict does not represent an anomaly, but rather a particular confluence of the same basic mechanisms that shape other types of human interaction. Indeed, the distinction between constructive and intractable conflict has parallels to distinctions between phenomena in the physical sciences. A hurricane, for example, is a rare event with destructive consequences, but it represents an expression of the same elements (air molecules, water) and forces (barometric pressure, wind currents, temperature) that promote sunny skies.

Chapter Overview

A central aim of this book is to couch intractable conflict—the form that poses serious challenges for social life—in terms of basic principles that find expression everywhere in the social and physical sciences. This chapter is intended to set the stage for this endeavor by presenting three seemingly different ways of understanding and investigating the potential for destructive conflict. The first perspective is the most straightforward: social science theory and research on interpersonal and intergroup conflict (cf. Deutsch, Coleman, & Marcus, 2006). By focusing directly on the experience of conflict in real-world settings, this approach has documented a plethora of local conditions and precipitating factors associated with the eruption and maintenance of antagonistic social relations. We then broaden our perspective by discussing the relevance of social psychology—the study of human thought and behavior in a social context. This approach assumes that conflict, even in its most destructive forms, is an expression of basic principles that can be observed in any social relationship.

After framing the potential for understanding intractable conflict in terms of generic social processes, we introduce the idea that the principles at work in social conflict are not limited to those in the social sciences or psychology, but rather reflect those that are common to the physical sciences as well (e.g., Coleman, Bui-Wrzosinska, Vallacher, & Nowak, 2006; Vallacher, Coleman, Nowak, & Bui-Wrosinska, 2010). This broadened perspective on conflict is possible because of recent advances in the understanding and investigation of *nonlinear dynamical systems* in mathematics and the physical sciences (cf. Holland, 1995a, 1995b;

Strogatz, 2003). It is this perspective that provides the theme for this book. We wish to emphasize at the outset, though, that this way of framing the issue of intractable conflict does not dismiss the insights and findings generated within the social science and social psychological perspectives. Rather, the perspective of nonlinear dynamical systems and complexity science provides a platform within which the propositions and data of conflict theory, peace research, and social psychology can be synthesized to create a unified account of one of the most pressing problems facing humanity today.

We conclude the chapter by providing succinct overviews of the succeeding chapters. The sequence of chapters is intended to be cumulative, with each chapter building on the ideas and research findings presented in the preceding chapters. Taken together, the chapters should provide a progressively integrated view of conflict, one that does justice both to the diversity of conflict in the real world and to the common principles that unite them all in theory.

The Problem of Intractable Conflict

Conflict is traditionally defined as *the perception of incompatible activities* (goals, claims, beliefs, values, wishes, actions, feelings, etc.). An incompatible activity "prevents, obstructs, interferes, injures" or in some way makes less likely or less effective another activity (Deutsch, 1973, p. 10). When a conflict is perceived, we actively interpret its meaning through pre-existing cognitive structures (beliefs, attitudes, stereotypes, etc.), through a consideration of the context of the conflict (occurring in the context of a long friendship or between enemies), and by way of certain mediums or processes (such as direct perceptions vs. second-hand reports or rumors; see Brunswick, 1956; Deutsch, 1973). At any point in this process of perception and interpretation, conflicts can begin to be seen as more or less important, threatening, and intractable. For example, the printing of political cartoons depicting the prophet Mohammed in newspapers in Europe is seen as a provocative gesture to some, and as a flagrant attack on Islam to others. The interpretation of these activities, regardless of the editors' real or stated intentions, is significantly affected by the perceiver's psychological schema for the conflict, social interactions with peers, the medium through which they learn of the events (The *New York Times* or *Al Jazeera*), and by the contextualization of the event within their own or their group's normative understanding of Muslim–Non-Muslim relations worldwide.

Our definition of social conflict builds on the thinking of Follet (1925/1973), Lewin (1936) and Deutsch (1973), but incorporates contemporary calls to view conflict not as a moment-in-time, but rather as a process unfolding in relationships across time (De Dreu, 2010; De Dreu & Gelfand, 2008; Pondy, 1967). Accordingly, we define conflict as *a relational process that is influenced by the perception of incompatible activities-difference*. These processes typically occur in a relational context that provides a sense of history and a normative trajectory. In other words, the perception of incompatible activities can function to redefine the manner in which the parties to a relationship think about and act toward one another.

Intractable Conflict in the Contemporary World

The news about global conflict is mixed. The International Crisis Group is currently monitoring 70 conflicts worldwide (as of today: 5 crisis, 9 deteriorating, 55 status quo, 1 improved). The good news is that the world has seen a sharp decrease in international conflicts since the end of the Cold War. Although there was a sharp increase in intra-national (civil) conflicts following the collapse of the Soviet empire, this trajectory peaked in 1991, with a 40 % decrease in the number of civil wars by 2003. We have also seen dramatic increases in wars ending through negotiation. Indeed, today twice as many conflicts end through negotiation than through military victory—a proportion that has flipped since the Cold War). To illustrate, from 1980–2003, more wars ended through negotiation than had ended in this fashion in the previous two centuries.

The bad news is that 25 % of wars ended through negotiation relapse into violence within 5 years. In some cases (such as Rwanda and Angola), more people were killed in-country after peace agreements failed than were killed during the war itself. States with civil wars, moreover, are far more likely to experience new violence. In fact, the longer a civil war lasts, the greater the likelihood of a recurrence of violence. In addition, civilian casualty rates have skyrocketed (today 80–90 %), and hospitals, schools, and children are directly targeted.

Highly destructive conflicts are often difficult to resolve. Of the 70 ongoing ICG-monitored conflicts noted above, over half (43) have persisted for 20 years or more (12 have persisted between 11 and 20 years, 15 have persisted between 1 and 10 years). History attests to the difficulty of resolving destructive conflicts. At the onset of WWI, none of the major powers expected a prolonged conflict. Four years later, 70 million troops had been mobilized and nine million combatants lay dead. Between 1945 and 1995, 18 cases of intractable interstate relationships have been documented, producing 75 militarized and violent conflicts that resisted hundreds of attempts at resolution, despite the threats they posed to regional or international security (Bercovitch, 2005). Indeed, enduring conflicts have been linked to 50 % of the interstate wars since 1816, with 10 out of 12 of the most severe international wars emerging from protracted destructive relations (Bennett, 1996). The apparent immunity to resolution has led many scholars to label such conflicts *intractable* (cf. Coleman, 2003). In view of the ubiquity and seeming intractability of destructive conflict in the world today, a clear vision of how such conflicts can be transformed and hopefully resolved constitutes what is arguably the major challenge of contemporary social science.

What We Know About Intractable Conflict

What makes a conflict intractable? A survey of the extant literature on intractable conflict presents a complex picture of the sources of intractability. In his meta-framework on intractable conflict, Coleman (2003) identified over 50 variables associated with the persistence of destructive conflicts. These include a variety of independent dimensions concerning the contexts, issues, relationships, processes, and outcomes of such conflicts. Coleman concluded that intractability is the result of complex interactions among multiple factors across different levels of these conflicts over long periods of time.

The centuries-old conflict in Northern Ireland is a good example of this multi-level complexity. The Irish "troubles," long understood as a religious conflict between Catholics and Protestants in Ireland, at its base is a conflict between those who wish to see Northern Ireland remain part of the United Kingdom, and those who wish to see the unification of the island of Ireland (Cairns & Darby, 1998). Religion, of course, plays a role in this conflict, as do global affairs, a history of international dominance, economic and other types of inequality (access to education, healthcare, housing, jobs, etc.), issues of social identity, the existence of multiple factions within each community, and serious concerns over human rights abuses and the use of terrorist tactics. These structural and group-level factors have a considerable impact on interpersonal relations (between friends and enemies) and personal functioning (mental and physical health, decision-making, voting behavior, etc.). Thus, long-term patterns of inter-ethnic violence in Northern Ireland are multiply determined.

Several lines of research are devoted to conflict intractability. Labels such as deeply-rooted conflict (Burton, 1987), protracted social conflict (Azar, 1986, 1990), moral conflict (Pearce & Littlejohn, 1997), and enduring rivalries (Goertz & Diehl, 1993) have been used to depict conflicts of this nature. Kriesberg (2005) stresses three dimensions that distinguish intractable from tractable conflicts: their persistence, destructiveness, and resistance to resolution. Most conflicts don't begin as intractable, but become so as escalation, negative sentiment, and hostile cognitions and interactions change the quality of the conflict. They can be triggered by a wide variety of factors and events, but often involve such issues as moral and identity differences, high-stakes resources, and/or struggles for power and self-determination (Coleman, 2003, 2006; Kriesberg, 1999). Not surprisingly, these circumstances often lead to incalculable human suffering, including destruction of vital infrastructure, division of families and communities, extreme violence, dislocation, and trauma (Cairns & Darby, 1998; Coleman, 2000a, 2000b).

What We Don't Know About Intractable Conflict

Despite the widespread and destructive nature of intractable conflict, this phenomenon has yet to be conceptualized in an agreed-upon and coherent fashion. The failure to achieve consensus regarding the fundamental processes underlying intractable conflict, and the corresponding failure to generate effective strategies for transforming such conflict, does not represent a lack of effort on the part of the scientific and practitioner communities. To the contrary, numerous theories, research initiatives, and intervention strategies have been proposed over the years (cf. beyondintractability.org; Azar, 1990; Burton, 1987; Cairns & Darby, 1998; Coleman, 2003, 2004, 2006; Crocker & Hampson, 2004; Crocker, Hampson, & Aall, 2005; Diehl & Goertz, 2001; Fiol, Pratt, & O'Connor, 2009; Goertz & Diehl, 1993; Kelman, 1999; Kriesberg, 1999, 2005; Kriesberg, Northrup, & Thorson, 1989; Lederach, 1997; Pearce & Littlejohn, 1997; Pruitt & Olczak, 1995). To some extent, the problem in framing a coherent theory reflects the inevitable idiosyncrasies of each conflict. Common factors and processes have been identified, but they represent an embarrassment of riches for theory construction. The challenge for theory construction is how to integrate these diverse factors into an account that is coherent, yet allows for prediction and a basis for conflict resolution in specific conflict settings.

A defining feature of intractable conflicts is that they display remarkable resistance to intervention even in the face of rational considerations that would seemingly defuse the animosities promoting the conflict (cf. Azar, 1990; Bar-Tal, 2007; Bennett, 1996; Bercovitch, 2005; Burton, 1987; Coleman, 2003; Goertz & Diehl, 1993; Kriesberg, 2005; Marshall & Gurr, 2005; Pearce & Littlejohn, 1997). In fact, there is some evidence that mediation has no impact at all in these situations or in fact makes these conflicts worse (Diehl & Goertz, 2001). This suggests that the problem of intractability says more about psychology than it does about objective reality. Numerous psychological mechanisms relevant to conflict intractability have in fact been identified (cf. Deutsch et al., 2006). Again, the challenge of achieving theoretical clarity does not reflect a lack of identifiable factors, but rather an overabundance. An intractable conflict is one that has become embedded in a host of cognitive, affective, and social-structural mechanisms, a transformation that effectively distances the conflict from the perceived incompatibilities that launched it.

How might we advance our understanding of such states of intractability? Can we better specify and model the underlying structure and dynamics of these phenomena? And can such a model accommodate the multitude of variables and processes identified as contributing to intractability, while remaining sufficiently parsimonious and predictive? We suggest that the perspective provided by nonlinear dynamical systems and complexity science offers a fruitful platform for the development of such a framework. Before introducing this perspective, however, it is useful to frame intractable conflict in terms of basic social processes that have been shown to characterize human experience generally.

The Relevance of Social Psychology

No area of social relations is untouched by the potential for conflict. Warmth, trust, humor, and empathy are very important dimensions of human interaction, but none of these come close to conflict in their ubiquity across all forms of social encounter. Conflict is manifest at all levels of social reality, from dyadic interactions to international relations. It takes place on an enormous range of time scales, from a minor dust-up at a social gathering to civil wars that persist for decades. It occurs in relationships with very different degrees of depth, from complete strangers to the most intimate lovers. And conflicts run the gamut from those that are constructive, such as parliamentary debate and group therapy, to those that are destructive and detrimental to all concerned, such as war and genocide.

Because conflict is so central to social life, the scientific study of social relations—social psychology—is ideally situated to identify the basic principles underlying the myriad manifestations of conflict. According to the field's first historian, F.B. Karpf, the devastation wrought by the U.S. Civil War motivated the early development of social psychology (Morawski, 2000). And indeed, conflict has been a recurring theme in social psychology since the field's beginnings in the early twentieth century. Kurt Lewin (1936), arguably the founding father of social psychology, devoted a great deal of attention to the study of conflict, analyzing it in terms of a field of psychological forces. Lewin's students, most notably Morton Deutsch, picked up on his lead and developed coherent and heuristic theories of conflict. Deutsch (1973), for example, framed conflict in terms of basic dimensions of social relations, such as cooperation versus competition, and distinguished between constructive and destructive conflicts.

The Advantages of Social Psychology

Social psychology came of age in the aftermath of World War II. Entire countries had lined up against one another and engaged in military campaigns that eventually led to the death of over ten million people, many of them victims of genocide in service of ethnic cleansing. This clearly was not the first time that nations had engaged in warfare; human history is littered with violent conflicts that erupt with disturbing regularity. Social psychology, however, had recently emerged as a scientific discipline with the methods—primarily those involving experimentation—to identify the factors that give rise to such events. Laboratories devoted to the study of conflict were established at several institutions in the United States, including Columbia University, University of California at Berkeley, Massachusetts Institute of Technology, and University of Michigan, and at various institutions in other countries as well. In its aspiration to become a respected science, social psychology turned increasingly to the experimental method, fashioned along the lines of methods employed in the physical sciences to identify causal mechanisms. This approach has served social psychology well. In the context of a well-designed experiment, one can isolate potential causes while controlling for others, and examine the effects of these "independent variables" on the phenomenon of interest, which is defined in terms of agreed-upon measurement procedures that reduce its ambiguity. The real world is complex and messy, with variables linked together by unclear causal relations, so bringing selected features of the world into the lab and systematically exploring their causal underpinnings brings a great deal of added value to the study of conflict. In identifying cause and effect, experiments have a crucial advantage over approaches that identify patterns of correlation among variables without being able to determine the mechanisms responsible for their association.

The Limitations of Social Psychology

The control and precision afforded by experimental methods, however, come at a cost. For one thing, experiments by their very nature are artificial and thus lack *mundane realism*—similarity to settings in real life. There is also the risk that they come up short on *psychological realism*—the ability to trigger the psychological processes that the experiment is designed to create. Social psychologists are keenly aware of these limitations of experimentation, of course, and have become very adept at creating realistic settings in laboratory settings, often by employing clever cover stories that deceive participants into believing things about their experience that are not true. Still, it is difficult to recreate the intensity and personal involvement inherent in real-world conflict in a psychology laboratory on a nice tree-lined college campus.

Experiments are also limited because certain topics of intense interest are taken off the table. All the informed consent and debriefing in the world would not justify testing the conditions under which people experience personal humiliation or engage in armed conflict with people from different social groups. Yet such issues are often front and center in real-world conflicts. Excluding them from consideration limits what can be learned from experimental research, and thus may provide an incomplete picture of how certain forms of conflict develop and become maintained.

Ironically, the focus on causality poses another limitation on the value of experimental research. In the prototypical experiment, one or more independent variables are manipulated at Time 1 and their main and interactive effects are assessed at Time 2. A psychological process, however, does not necessarily stop after a cause has produced an effect. Indeed, the immediate effect can itself function as a cause, changing the course of subsequent behavior, perhaps even magnifying or diminishing the very causal factor that produced the effect in the first place. Experimental research has established that exposure to televised violence promotes aggressive behavior in young children, for example, but this conclusion creates the impression that the relationship is a one-way street. The real world is awash in reciprocal causality, however, with the initial effect of a causal factor functioning as a cause as the process unfolds. Watching a violent TV show may indeed lower a child's threshold for aggression, but once the child behaves violently, he or she may be inclined to absorb even more violent TV shows.

The potential for reciprocal causality and other temporal patterns is especially acute in conflict scenarios that unfold over extended periods of time. The first response to an instigating factor may be important, but stopping the investigation at this point may provide a misleading portrait of the scenario. Over time, the initial response may become amplified in intensity, diminished in intensity, or follow a more complex time course such as periodic oscillation or chaos. Investigating conflict as a one-step process clearly does not do justice to the dynamics involved.

Experiments have another limitation that can prove even trickier to resolve in an unequivocal fashion. If the situation created in a laboratory setting is truly realistic for participants, it has the potential to alter the state of mind that participants had before coming to the experimental session. Some experiments are designed to make participants feel uncomfortable or to behave in ways they might not otherwise consider. Particularly in research relevant to social conflict, the need for psychological realism often necessitates placing participants in personally distressing situations. The famous experiments by Stanley Milgram (1974) on obedience to authority illustrate this concern. Psychologists were shocked by Milgram's results, which showed that a majority of participants followed the orders of the experimenter to deliver up to 450 V of electricity to a middle-aged man with a heart condition when he provided incorrect answers to a series of questions.

No doubt this experience was uncomfortable for participants, who may have realized for the first time that they are capable of inhumane behavior toward someone who did not warrant such treatment. Participants were debriefed upon completion of the session, of course, but telling them the true intent of the experiment may have made matters even worse. Imagine how you would feel if you delivered what you knew were painful shocks to someone else, only to learn afterward that your behavior had nothing to do with learning, but everything to do with blind obedience to authority.

Putting Social Psychology in Context

Social psychology experiments play a crucial role in understanding conflict because they identify causal factors and reveal how a small set of them interact to influence thoughts, feelings, and actions in conflict-relevant settings. By themselves, however, experiments cannot accommodate the enormous complexity of real-world conflicts, each of which is nested in an idiosyncratic ensemble of historical, geographical, economic, and social factors. An experiment is ideally suited for isolating specific causes, but is not well suited for probing the interactions among dozens of factors.

The benefits of social psychology are best realized in the context of multi-disciplinary research. The perspectives and methods of political scientists, anthropologists, economists, and historians provide insight into the complex patterns of intertwined factors at work in the emergence, maintenance, and resolution of conflicts between social entities (e.g., groups, cultures, nations). The role of social psychology in this collaborative strategy is to focus on the causal connections among subsets of the factors identified by their social science colleagues. The cross fertilization of theories and methods between social psychology and the other human sciences can promote a synthesis that captures both the mechanisms of conflict and the larger context in which these mechanisms operate. If performed competently, this multi-disciplinary approach can avoid the trade-off between precision and qualitative understanding associated with employing one approach and ignoring the others.

The value of multi-disciplinary research concerning conflict can, in effect, be *too* successful. Because conflicts represent the complex interactions among myriad factors at different levels of psychological and social reality, models that capture this complexity can be correspondingly complex and thus fail to provide coherent and parsimonious understanding of the nature of conflict.

This is where the paradigm of complexity science and dynamical systems enters the picture. This paradigm has revolutionized the physical sciences (cf. Holland, 1995a, 1995b; Strogatz, 2003; Waldrop, 1992), and in recent years it has been applied to important psychological and social processes as well (cf. Guastello, Koopmans, & Pincus, 2009; Nowak & Vallacher, 1998; Vallacher & Nowak, 1994a, 2007). As detailed below, the dynamical approach is designed to uncover the underlying forces that give rise to the complexity and richness of social relations. This approach, moreover, can track the operation of these forces as they interact with one another on different time scales. Once these forces and temporal trajectories are identified, experimental methods can be employed to isolate specific causal mechanisms and to test hypotheses concerning mechanisms that have yet to be identified. This approach to science thus has the potential to develop comprehensive models that recognize the multi-faceted nature of conflicts in the real world, but to do so with respect to basic principles that are both parsimonious and generalizable across manifestly different conflict scenarios.

The Relevance of Complexity Science

Everyone would agree that the world, in both its physical and social manifestations, is very complex, with a wide variety of forces and mechanisms responsible for the surface structure of reality. Any phenomenon—whether the formation of a galaxy, an internal combustion engine, brain function, or stock market patterns—can be conceptualized as a complex system composed of many parts that give rise to the behavior of interest. From its inception over 400 years ago, science has tried to understand complicated systems by reducing them into their simple parts. This eminently reasonable strategy has had its share of successes, but ultimately it falls short in providing insight into how the system as a whole manages to function. For many topics of interest to scientists and lay people, knowing the list of parts that make up a system and how these parts work in isolation simply does not provide insight into the strange and exciting things that happen when all those parts work together. Knowing all the carbon, nitrogen, oxygen, sodium, and chloride atoms in a teenager, for example, does not explain, let alone predict that crazy thing that he or she did last week.

This is also true for the conflicts between people, cultures, and nations. Knowing the parts of these systems does not necessarily inform us how they interact to generate heated conflicts, nor does such knowledge point to the interventions that may be needed to transform antagonistic relationships into relationships characterized by tolerance and good will.

The Essence of Complexity Science

Complexity science is a new scientific field that can give us some understanding of how the properties of whole systems arise from the interactions of their parts. The ideas that underlie this new science are based on the behavior of some example systems that are described by detailed mathematics. These examples are called "dynamical systems" because the variables that describe the state of the system are dynamic—they evolve over time. Mathematical, physical, and social complex dynamical systems share some important broad characteristics that are useful for us in understanding the social psychology of conflicts. The following characteristics are especially relevant to our depiction of conflicts that become intractable—protracted over time and seemingly impervious to resolution:

- Self-Organization. Local interactions can create large-scale patterns. The movement of tiny patches of moist hot air forms a hurricane extending over hundreds of miles. Actions of individual investors create economic bubbles and then burst them. The decisions by a few local Liberian mothers and grandmothers to employ non-violent forms of anti-war civil-disobedience result in the downfall of the strongman Charles Taylor and the emergence of peace in Liberia.
- Emergence. The properties of the whole system are often quite different from the properties of its parts. This is widely recognized in the physical sciences. For example, hydrogen and oxygen together are an explosive mixture of gasses, but water—which represents the interaction of hydrogen and oxygen—is stable and wet. Examples of emergence abound as well in the social sciences. For example, individually peaceful people can assemble into a dangerous, violent mob.
- Unintended Consequences. Small tinkering with, or changing the pieces of a system can lead to surprising and completely unanticipated results. A tree falls on an electrical transmission wire in a forest in the U. S. Midwest and cascading electrical failures put out the lights of tens of million of people in the Northeast.

The Internet, originally designed to transfer data files between military computers, leads to on-line social networks that mobilize average citizens into toppling a dictatorship.

Complexity Science and Social Conflict

This new science of complexity can be used in two different ways to help us understand social systems and conflict. In one approach, we can encode social mechanisms into equations, solve those equations (analytically or numerically) by rigorous mathematical methods, and thereby learn the logically necessary consequences of those social mechanisms. This approach, which is the one typically used in the physical sciences, is now being used in the social sciences to gain new insight into social phenomena. Examples of this approach later in this book include models of conflicts between two people and the rapid spread of conflict from an isolated region to an entire society.

We can also use our knowledge gained from these mathematical examples in an entirely different way. It is hard to think of entirely new things. Can you picture a color that you have never seen before? In science, we often use an idea from something that is familiar to us and apply it to something new. That approach can give us a new perspective and therefore new insights. In this way, we can use what we know about mathematical systems to give us new metaphors to understand social systems. Networks, sand hills, self-organization—these new metaphors help us see conflicts in a new way. They yield new interpretations of the data, they drive us to ask new questions, and they suggest new social psychology experiments for us to perform. Examples of this approach later in this book include understanding the intractability of conflicts, analysis of emotionally charged difficult conversations between people, and the sudden transformation from conflict to peace is societies ravaged for years by civil war.

Throughout this book we will use both mathematics and metaphors derived from mathematics based on several different types of complex dynamical systems. Here, we mention only one particular metaphor because we will use it extensively throughout this book. The metaphor, based on ordinary differential equations, consists of an artificial landscape of hills and valleys. The location in that landscape represents some characteristic of the conflict, for example, the emotional states of the participants. An artificial gravity relentlessly pulls those participants downhill into a valley. This valley is called an "attractor" because it represents the stable endpoint that results from the mutual behavior of the participants. Small changes may lead the participants uphill a little, but they are likely to soon fall back into their previous patterns of behavior represented by the bottom of the valley. Escape out of the valley, and therefore out of the conflict, is possible only if additional forces or a change in behavior get the participants past the top ridge that forms the valley, or a deeper more favorable valley opens up within their reach.

The Organization and Trajectory of This Book

The perspective we present in this book can itself be viewed in terms of complexity and dynamical systems. It is complex in that it is comprised of many distinct principles and provides integration for many different topics. Individually, the various principles provide a piece of the puzzle; collectively, they offer insight into the basic psychological and social functions that are common to levels of experience as distinct as emotion, social judgment, and intergroup relations. The perspective is dynamical in that the principles build upon one another to generate a coherent account of how social relations can evolve toward destructive conflict—and how they can reverse course and promote peace, tolerance, and goodwill.

With this reflexive approach to theory depiction in mind, we have disassembled our account into a set of specific ideas, each providing the theme for a distinct chapter. The chapters are then presented in an order intended to show how the principles are connected and provide progressive insight into the nature of intractable conflicts and the means by which they can be resolved. An overview of this organization and trajectory is provided in the following chapter-by-chapter synopses. Of course, each synopsis is a synthesis of yet more basic points; these will be unpacked in accordance with an appropriate trajectory in their respective chapters.

Chapter 2. *Origins: The Promise of Dynamical Systems Theory*

Scholarly and practical interest in conflict has evolved from a focus on peace and conflict *per se* to a focus on basic social processes that can give rise to hostile interpersonal and inter-group relations under certain conditions and in response to various factors. Indeed, conflict is often a constructive phenomenon, setting the stage for new forms of thought and action, and promoting a shared reality among individuals and groups. A small proportion of conflicts, however, do not conform to canonical models of human motivation emphasizing enlightened self-interest, rationality, or even hedonism. Conflict, moreover, is manifest in many different ways and at different levels of social reality. The paradoxical nature and diversity of conflict poses a serious challenge for comprehensive theory construction. Recent developments in complexity science and nonlinear dynamical systems, however, have transformed our understanding of phenomena in the physical world, and there is reason to believe that this approach can promote a coherent view of conflict in all its guises.

Chapter 3. *Foundations: The Dynamical Perspective on Social Processes*

An appreciation for the complex and dynamic nature of social processes can be traced to the pioneers of social psychology in the early twentieth century. In recent years, these early insights have found expression in theory and research through the adaptation of the concepts, principles, and tools of nonlinear dynamical systems. Although relatively new, this approach—*dynamical social psychology*—is showing promise in integrating different domains of social experience with respect to basic dynamic processes that are observed in phenomena throughout nature. Rather than focusing on linear cause-effect relations, the dynamical approach emphasizes the emergence of higher order states through the self-organization of a system's basic elements. These emergent properties, in turn, provide coherence for the system in question—from self-concepts to international relations—and resist change due to external influence. This basic scenario provides an integrative platform for investigating the link between structure and dynamics at different levels of social reality—from minds to societies. In so doing, the dynamical perspective sets the stage for reframing the essential dynamics of conflict.

Chapter 4. *Patterns: Trajectories of Conflict*

Traditional approaches to social conflict focus on central tendencies, such as the average amount of distrust, prejudice, or violence between mutually antagonistic individuals or groups. In the dynamical approach, the focus is on *patterns* of cognition, affect, and behavior that lose their meaning if averaged over time and circumstances. The tension between warring parties may show periodic or quasi-periodic oscillation on embedded time scales, for example, or display a seemingly random temporal pattern that reflects deterministic chaos. Interpersonal and inter-group relations are also characterized by coherent higher-order states that function as *attractors* for cognition, affect, and behavior. The nature of these attractors, in combination with outside influences on the system, shapes the temporal pattern of conflict. If governed by weak attractors, conflict in a relationship displays a linear pattern, with each party responding in a proportional manner to provocations and conciliatory gestures. If governed by strong attractors, conflict in a relationship displays a nonlinear pattern characterized by sustained periods of little overt change punctuated by sudden and dramatic changes in the conflict's intensity.

Chapter 5. *Traps: Intractable Conflict as a Dynamical System*

Most conflicts are resolved fairly quickly and constructively, but a small proportion of them become protracted and resistant to positive transformation. These *intractable conflicts* reflect a breakdown in self-regulation and a loss of complexity as a strong destructive attractor comes to constrain the pattern of thought and behavior in the relationship, with more constructive attractors dissipating or becoming latent. In this process, the conflict may spread to all spheres of interaction among the entities, making it increasingly difficult to resolve. Indeed, the event or circumstance that initially launched the conflict may be resolved (or become irrelevant), yet the conflict will be maintained. This scenario of self-sustaining and increasingly pervasive conflict has been documented empirically in social interactions and modeled in computer simulations of societal dynamics. The chapter concludes with some thoughts about the closed nature of intractable conflict systems, and potential methods for modeling and investigating open versus closed system dynamics.

Chapter 6. *Escape: How Intractable Conflicts Can Be Transformed*

Traditional approaches to conflict resolution tend to focus on basic cause-effect relations operating on relatively short time scales. By overlooking the patterns and potential for emergence in a destructive relationship, the emphasis on "one-step" causation is not well suited for changing the dynamics sustaining the conflict. In contrast, reconfiguring the dynamics of interpersonal, intergroup, or international relations is the focus of the dynamical approach to transforming intractable conflicts. There are three ways of accomplishing this transformation. *Attractor deconstruction* entails restoring complexity to a conflict scenario by isolating elements or changing the feedback loops among them. The *creation of a latent attractor* trades on the tendency toward multi-stability in systems and entails the consolidation of incongruent (positive) elements into a coherent structure. In the *bifurcation scenario*, factors are identified that can change the number and types of attractors in a conflict scenario. Multi-disciplinary research provides preliminary support for these approaches.

Chapter 7. *Sustainability: The Dynamics of Enduring Peace*

Peace is more than the absence of destructive conflict. Attempts to suppress conflict can, in fact, have the ironic effect of promoting catastrophic ruptures in interpersonal, intergroup, and international relations. Nor is peace a stable state. Rather, it evolves in the context of ongoing social relations. Sustainable peace must therefore be approached with an appreciation of dynamic processes, including the development and maintenance of attractors for social relations, as well the recognition that attractors for conflict and violence may remain latent in the system. Insight into the dynamics of sustainable peace can generate specific strategies of peacemaking and peacekeeping. Case studies of successful transformation from war to peace demonstrate the efficacy of these strategies.

Chapter 8. *Epilogue: Conflict in the Twenty First Century*

Conflict occurs in the context of global dynamics. Because of climate change, energy crises, and immigration patterns, competition over scarce resources such as oil, water, and land amplifies the potential for destructive conflict. Because of increasing contact and interdependence among groups with different worldviews and religions, the future may see these resource conflicts devolve into identity-based conflicts. And because of modernity, with its practical and existential consequences, attractors for many current forms of social relations will be destabilized. But there is also reason for optimism as the twenty first century unfolds. Indeed, many of the factors that paint a pessimistic scenario—scarce resources, globalization and intergroup contact, exposure to different worldviews—may foster mutual respect and compromise rather than disdain and hostility. This potential for bifurcation—with the same set of factors capable of promoting the emergence of very different higher-order states—is a hallmark of nonlinear dynamical systems. Which scenario will prevail will depend on how we manage the dynamics of an increasingly complex and interconnected planet.

References

Azar, E. E. (1986). Management of protracted social conflict in the third world. *Ethnic Studies Report, 4*, 1–17.

Azar, E. E. (1990). *The management of protracted social conflict*. Aldershot: Dartmouth.

Bar-Tal, D. (2007). Social-psychological foundations of intractable conflict. *The American Behavioral Scientist, 50*, 1430–1453.

Bennett, D. S. (1996). Security, bargaining, and the end of interstate rivalry. *International Studies Quarterly, 40*, 157–184.

Bercovitch, J. (2005). Mediation in the most resistant cases. In C. A. Crocker, F. O. Hampson, & P. Aall (Eds.), *Grasping the nettle: Analyzing cases of intractable conflict*. Washington, DC: United States Institute of Peace.

Brunswick, E. (1956). *Perception and the representative design of experiments* (2nd ed.). Berkeley: University of California Press.

Burton, J. (1987). *Resolving deep-rooted conflict: A handbook*. Lanham, MD: University Press of America.

Cairns, E., & Darby, J. (1998). The conflict in Northern Ireland: Causes, consequences, and controls. *The American Psychologist, 53*, 754–760.

Coleman, P. T. (2000a). Intractable conflict. In M. Deutsch & P. T. Coleman (Eds.), *The handbook of conflict resolution: Theory and practice* (pp. 428–450). San Francisco, CA: Jossey Bass.
Coleman, P. T. (2000b). Fostering ripeness in seemingly intractable conflict: An experimental study. *International Journal of Conflict Management, 11*, 300–317.
Coleman, P. T. (2003). Characteristics of protracted, intractable conflict: Toward the development of a metaframework—I. *Peace and Conflict: Journal of Peace Psychology, 9*, 1–37.
Coleman, P. T. (2004). Paradigmatic framing of protracted, intractable conflict: Towards the development of a meta-framework—II. *Peace and Conflict: Journal of Peace Psychology, 10*, 197–235.
Coleman, P. T. (2006). Conflict, complexity, and change: A meta-framework for addressing protracted, intractable conflicts—III. *Peace and Conflict: Journal of Peace Psychology, 12*, 325–348.
Coleman, P. T. (2011). *The five percent: Finding solutions to seemingly impossible conflicts*. New York, NY: Public Affairs.
Coleman, P. T., Bui-Wrzosinska, L., Vallacher, R. R., & Nowak, A. (2006). Protracted conflicts as dynamical systems. In A. K. Schneider & C. Honeyman (Eds.), *The negotiator's fieldbook: The desk reference for the experienced negotiator* (pp. 61–74). Chicago: American Bar Association.
Crocker, C. A., & Hampson, F. O. (2004). *Taming intractable conflicts: Mediation in the hardest cases*. Washington, DC: United States Institute of Peace.
Crocker, C. A., Hampson, F. O., & Aall, P. (2005). *Grasping the nettle: Analyzing cases of intractable conflict*. Washington, DC: United States Institute of Peace.
De Dreu, C. K. W. (2010). Social conflict: The emergence and consequences of struggle and negotiation. In S. T. Fiske, D. T. Gilbert, & G. Lindzey (Eds.), *Handbook of social psychology* (5th ed., pp. 983–1023). San Francisco, CA: Wiley.
De Dreu, C. K. W., & Gelfand, M. J. (2008). Conflict in the workplace: Sources, dynamics and functions across multiple levels of analysis. In C. K. W. De Dreu & M. J. Gelfand (Eds.), *The psychology of conflict and conflict management in organizations*. New York, NY: Lawrence Erlbaum.
Deutsch, M. (1973). *The resolution of conflict: Constructive and destructive processes*. New Haven, CT: Yale University Press.
Deutsch, M., Coleman, P. T., & Marcus, E. C. (Eds.). (2006). *Handbook of conflict resolution: Theory and practice* (2nd ed.). San Francisco, CA: Jossey-Bass.
Diehl, P. F., & Goertz, G. (2001). *War and peace in international rivalry*. Ann Arbor, MI: University of Michigan Press.
Fiol, C. M., Pratt, M. G., & O'Connor, E. J. (2009). Managing intractable identity conflicts. *The Academy of Management Review, 34*, 32–55.
Follet, M. P. (1973). Power. In E. M. Fox & L. Urwick (Eds.), *Dynamic administration: The collected papers of Mary Parker Follett* (pp. 66–87). London: Pitman. Original work published in 1925.
Goertz, G., & Diehl, P. F. (1993). Enduring rivalries: Theoretical constructs and empirical patterns. *International Studies Quarterly, 37*, 147–171.
Guastello, S., Koopmans, M., & Pincus, D. (Eds.). (2009). *Chaos and complexity in psychology: The theory of nonlinear dynamical systems*. New York, NY: Cambridge University Press.
Holland, J. (1995a). Can there be a unified theory of complex adaptive systems? In H. J. Morowitz & J. L. Singer (Eds.), *The mind, the brain, and complex adaptive systems*. Boston, MA: Addison-Wesley.
Holland, J. H. (1995b). *Emergence: From chaos to order*. Reading, MA: Addison-Wesley.
Kelman, H. (1999). The interdependence of Israeli and Palestinian national identities: The role of the other in existential conflicts. *Journal of Social Issues, 55*, 581–600.
Kriesberg, L. (1999). Intractable conflict. In E. Weiner (Ed.), *The handbook of interethnic coexistence* (pp. 332–342). New York, NY: Continuum.

Kriesberg, L. (2005). Nature, dynamics, and phases of intractability. In C. A. Crocker, F. O. Hampson, & P. Aall (Eds.), *Grasping the nettle: Analyzing cases of intractable conflict* (pp. 65–98). Washington, DC: United States Institute of Peace.
Kriesberg, L., Northrup, T. A., & Thorson, S. J. (1989). *Intractable conflicts and their transformation*. Syracuse, NY: Syracuse University Press.
Lederach, J. P. (1997). *Building peace: Sustainable reconciliation in divided societies*. Washington, DC: United States Institute of Peace.
Lewin, K. (1936). *Principles of topological psychology*. New York, NY: McGraw-Hill.
Marshall, M. G., & Gurr, T. R. (2005). *Peace and conflict*. College Park, MD: Center for International Development and Conflict Management.
Milgram, S. (1974). *Obedience to authority: An experimental view*. New York, NY: Harper & Row.
Morawski, J. (2000). Social psychology a century ago. *The American Psychologist, 55*, 427–430.
Nowak, A., & Vallacher, R. (1998). *Dynamical social psychology*. New York, NY: Guilford Press.
Pearce, W. B., & Littlejohn, S. W. (1997). *Moral conflict: When social worlds collide*. Thousand Oaks, CA: Sage.
Pondy, L. R. (1967). Organizational conflict: Concepts and models. *Administrative Science Quarterly, 12*, 296–320.
Pruitt, D. G., & Olczak, P. V. (1995). Resolving seemingly intractable conflict. In B. B. Bunker & J. Z. Rubin (Eds.), *Conflict, cooperation, and justice: Essays inspired by the work of Morton Deutsch*. San Francisco, CA: Jossey-Bass.
Strogatz, S. (2003). *Sync: The emerging science of spontaneous order*. New York, NY: Hyperion Books.
Vallacher, R. R., Coleman, P. T., Nowak, A., & Bui-Wrosinska, L. (2010). Rethinking intractable conflict: The perspective of dynamical systems. *The American Psychologist, 65*, 262–278.
Vallacher, R. R., & Nowak, A. (Eds.). (1994a). *Dynamical systems in social psychology*. San Diego, CA: Academic.
Vallacher, R. R., & Nowak, A. (2007). Dynamical social psychology: Finding order in the flow of human experience. In A. W. Kruglanski & E. T. Higgins (Eds.), *Social psychology: Handbook of basic principles* (pp. 734–758). New York, NY: Guilford.
Waldrop, M. M. (1992). *Complexity: The emerging science at the edge of order and chaos*. New York, NY: Schuster.

Chapter 2
Origins: The Promise of Dynamical Systems Theory

> *"Theory is an arbitrary structure that we impose on chaos to make it meaningful and predictable"*
>
> John Whiting

The dynamical systems approach to conflict is relatively new, but it has deep roots in other orientations and research agendas. Particularly noteworthy are three very distinct areas of inquiry with equally distinct historical pedigrees: peace and conflict studies, social psychology, and complexity science. As the John Whiting quote implies, each of these traditions is valuable yet limited, focusing attention on particular aspects of the phenomenon, often at the expense of others. Considered together, however, these three lenses impose structure on the "chaos" of conflict, enabling the emergence of a unique and coherent perspective on the development, maintenance, and resolution of conflict in interpersonal, intergroup, and international relations.

Of course, a comprehensive rendering of the impact these three perspectives have had on the study of social conflict would require a volume in its own right. Our aim here is to highlight those ideas and findings from each that have most informed our thinking. We first discuss the direct focus on conflict that has been provided by the peace and conflict perspective. Decades of theory, research, and practice have revealed paradoxical features of conflict and established the diverse forms that conflict can take at different levels of social reality. Within this tradition, five paradigms have been advanced to provide insight into the principles by which conflicts develop and unfold. In the next section, we discuss how conflict fits into general theories of social psychological functioning. We briefly outline six classic social psychological perspectives on conflict and then discuss basic issues of conflict in light of themes and questions that are ascendant in contemporary social psychology. In the last section, we turn our attention to recent developments in complexity science and dynamical systems, and the potential value these developments have for the study of conflict.

R.R. Vallacher et al., *Attracted to Conflict: Dynamic Foundations of Destructive Social Relations*, Peace Psychology Book Series, DOI 10.1007/978-3-642-35280-5_2,

Lessons from Peace and Conflict Studies

The eclectic field of *peace and conflict studies* has provided a wide variety of metaphors through which we can envision and study the phenomenon of conflict. In different quarters, scholars and practitioners have framed conflict in terms of *games*, *problems*, *diseases*, *force fields*, *machines*, *wars*, and *quagmires*. Each metaphor has value, but we prefer the metaphor offered by the eminent peace psychologist, Morton Deutsch, who is fond of saying "conflict is like sex". Like sex, conflict is everywhere. It is a natural state of affairs in the human condition, basic to life, learning, and love. It can be invigorating and deeply satisfying or hurtful and humiliating. It can be fleeting and insignificant or intense and life altering. Its lure can subside and remain latent for long periods of time, and then suddenly re-emerge and become the main focus of life. It can take place when we are alone, with other people, or with groups of people. The pathologies of conflict are similar to those of sex: people can become extremely conflict-avoidant or conflict-obsessed, suffer from position-rigidity or premature conflict resolution, and engage in it with extreme passion and emotionality or in an overly intellectualized and game-like manner. But ideally, conflict, like sex, results in satisfying experiences for all parties involved and enhances, not impairs, relationships.

This, then, begs the question that has organized over 70 years of systematic research on conflict: *What determines whether conflict—a pervasive and naturally occurring event—moves in a constructive or destructive direction*? (see Deutsch, 1973, 2006). In other words, what are the conditions and processes that determine whether conflicts provide a uniquely human opportunity to learn about ourselves and others, to motivate necessary changes in the status quo, to challenge obsolete ways of thinking, relating, working, and to innovate—or result in experiences of loss, frustration or misery. *Constructive conflicts* are defined as those which result in mutually satisfactory experiences of the processes, relationships and outcomes associated with the conflict for all involved parties (Deutsch, 1973). *Destructive conflicts* are the opposite, in them at least one of the parties involved experiences dissatisfaction with the conflict. Many if not most of the conflicts that we face have the potential for satisfying, constructive outcomes for all. However, this potential is rarely realized because of our tendency to see most conflicts as win-lose. Engaging in conflict tends to generate anxiety in many people who associate it with negative or violent outcomes, which leads to fight or flight responses. Good cooperative relations however facilitate the constructive management of conflict and the ability to handle constructively the inevitable conflicts that occur during cooperation, which facilitates the survival and deepening of cooperative relations.

The Paradoxes of Social Conflict

The view of conflict emanating from peace and conflict studies research is one of good news and bad news. The good news is that most conflicts are either fleeting and insignificant or are resolved constructively—or at least well enough. In fact, the historical trends are very promising. Violent conflict in Western society shows a clear downward trend throughout recorded history (Gurr & Scarritt, 1989; Pinker, 2011). For instance, the international community has recently experienced a dramatic increase in the number of wars ending through negotiation rather than through unilateral military victory. These numbers have flipped since the end of the Cold War, with today double the percentage of wars ending through negotiations than had previously through military victory (Mason, Crenshaw, McClintock, & Walter, 2007). Incredibly, more wars ended through negotiations from 1988 to 2003 than during the previous two centuries ((United Nations, 2004). After peaking in 1991, the number of civil wars dropped roughly 40 % by 2003 (United Nations, 2004). This suggests that local, regional, and international peacemakers are having an increasingly positive impact on constructive resolution.

But some conflicts are different—they are at odds with these positive trends. They act strangely and do not conform to normal expectations or respond to standard interventions. They resist change, grinding on and on for months, years, decades, even generations. Although these types of conflicts are relatively uncommon, they can cause disproportionate amounts of misery and instability, wreaking havoc on families, communities, nations, and at times entire regions of the globe (Coleman, 2003, 2011; Diehl & Goertz, 2001; Kreisberg, 2005). These conflicts have come to be labeled *intractable*.

Intractable conflicts evidence certain peculiarities. For example, they defy what we know about motivation and conflict. Research on *ripeness theory* has found that when people engaged in long-term conflicts experience high levels of pain and suffering from the conflict and then see some opportunity to escape it, they will tend to agree to negotiations and settlement (Druckman, 2007; Mooradian & Druckman, 1999; Zartman, 2000). However, sometimes the opposite occurs. Zartman (2000) has found that when conflicts involve "true believers" and "true believer cultures," extreme suffering can have the reverse effect. He writes, "Parties thinking as true believers are unlikely to be led to compromise by increased pain; instead, pain is likely to justify renewed struggle" (p. 239).

Intractable conflicts also display unique patterns of escalation and de-escalation. They may simmer at low levels of intensity for long periods of time, but then escalate into violence and bloodshed overnight, seemingly out of nowhere, showing spikes of enmity and hostilities previously unseen. These high-levels of intensity may persist much longer than anticipated, outliving any rational utility. But then one day, they may decline precipitously in violence—again, seemingly out of nowhere and for no obvious reason.

These types of conflicts tend to respond to direct attempts at intervention such as conciliatory gestures, mediation, or even threats of violence in odd and unpredictable ways. Specifically, they may (1) completely ignore the intervention, continuing their trajectory of escalation; (2) show an exaggerated response to a seemingly insignificant intervention, though in the desired direction; (3) evolve in a completely unpredictable direction independent of the intervention; or (4) respond in a manner proportional to the actions and plans undertaken by the intervener. Furthermore, intractable conflicts often respond to initiatives in ways that are disproportionate to the magnitude of intervention. In these situations, major attempts at resolution by superpowers, the UN, the international community, and other high-impact actors seem to have little or no observable effects on the conflict (Diehl & Goertz, 2001).

However, much smaller, seemingly insignificant actions—the involvement of a low-power group of clergy, women, or an NGO, or even a small gesture of genuine kindness by one of the parties—can have a major impact. This is what was witnessed in the protracted conflicts in both Liberia and Mozambique (Bartoli, Bui-Wrzosinska, & Nowak, 2010; Disney & Gbowee, 2012). Actions that seem small and insignificant often act as catalysts, facilitating processes of change when the critical ingredients for change are already in place. The critical ingredients for positive change center on social capital, collective identity, and a sense of hope. The critical ingredients for negative change, meanwhile, center on despair or a sense of hopelessness. Further, constituent groups at different levels of societies might respond to interventions in disparate ways. The experience of intractable conflict might for example lead the political elite to favor a negotiated settlement, but the population, having suffered so much pain, displacement, and grievance, might advocate continued struggle. Also, intractable societal conflicts can appear resolved through peace agreements, but the deep rooted cleavages in perceptions, grievances, etc. of ordinary people may not be addressed by elite level negotiations. Because of the difference in perception between societal layers, conflict can simmer on and re-erupt later.

Most strikingly, intractable conflicts seem to take on a life of their own. They can persist despite the weariness of the parties and an increasing genuine desire to end hostilities. And even when major changes occur in the political landscape—assassinations, coups, ousted leaders, policy changes, you name it—they can still persist. In this way they seem to defy logic and reason.

In sum, most conflicts have both positive and negative potential. They can propel people and societies forward toward new solutions, innovations, more just social arrangements and enhanced relationships—or they can stop these constructive developments dead in their tracks. The good news is that most conflicts are resolved effectively and we seem to be getting better at this. The bad news is when they go bad, they can get stuck and do an inordinate amount of damage.

But why? What is it about this particularly intractable species of conflict that makes them so odd, unpredictable, and paradoxical?

The Diversity of Conflict

Conflict is easy to recognize but hard to understand. It is often obvious and pervasive in human relations, yet it is like a Rorschach test for theorists, eliciting our own biases, fears, preferences and yearnings. As stated in Chap. 1, we define conflict as *a relational process that is influenced by the perception of incompatible activities*. These perceptions can occur within and between people and groups of people, can be expressed or left unexpressed, and can be experienced by the parties to the conflict or by observers external to the conflict (Boardman & Horowitz, 1994; Deutsch, 1973). These experiences can also differ by level of importance (from superficial to existential concerns), centrality (impacting one's identity, esteem, and sense of reality), pervasiveness (number of interconnections with other conflicts and experiences), and duration. Conflicts also contain important differences in the objective structures of the various issues involved. They may be purely distributive in nature (zero-sum, but divisible), integrative (with satisfactory alternatives available for all), inefficient (complex and difficult with elusive but nevertheless real integrative or distributive potential), or non-negotiable (with indivisible qualities). However, it is principally the *subjective experience* of conflict, which may or may not be congruent with the objective nature of the issues (see Deutsch, 1973), that drives reactions and behavior (Rubin, Pruitt, & Kim, 1994), particularly under conditions of intense escalation (Fisher & Keashly, 1990).

One set of challenges inherent to understanding intractable conflict concerns the complex and idiosyncratic nature of these conflicts, and the difficulty of generalizing knowledge across levels of analysis. Conflicts may be or may become intractable for a wide variety of reasons. In our research, we have identified over 50 aspects of such conflicts that scholars have recognized as accounting for intractability (Coleman, 2003, 2011; see Table 2.1). These includes various aspects of the contexts in which conflict occurs, the issues involved, the relationships that shape these issues, the processes that unfold, and the outcomes they generate. Every seemingly intractable conflict—Kashmir, Israel-Palestine, Northern Ireland, Colombia, etc.—is quite different, involving and emphasizing a unique subset of these factors. Indeed, the challenge of achieving theoretical clarity does not reflect a lack of identifiable factors, but rather an over-abundance. An intractable conflict is often one that has become entrenched in a wide variety of cognitive, affective, and social-structural mechanisms, a transformation that effectively distances the conflict from the perceived incompatibilities that launched it initially. Hence, the task for theory construction is integrating these diverse factors into an account that provides a coherent perspective, yet allows for prediction and a basis for conflict resolution in specific settings.

Further complicating theoretical comprehension of these protracted disputes is the fact that they occur at all levels of social relations, from marriages and families to civil wars and interstate warfare. Therefore, it may not always be useful or valid to compare, say, moral interpersonal conflicts with intractable conflicts over territory or water rights, or conflicts between a husband and wife in the U.S. with those

Table 2.1 Fifty-seven elements of intractable conflict

Context	Historical dominance and injustice	Domination	Severe imbalance of power between people or groups
		Inequity	History of colonialism, racism, sexism, ethnocentrism, or human rights abuses
		Gendered problems	Situations where men, who are responsible for the vast majority of violence, are in charge.
		Divide and conquer	High-power groups (HPGs) manipulate low-power groups' (LPGs') ethnic differences
		Cracks in the façade	Conditions where HPGs control of historical and cultural meaning through history textbooks, media, official accounts etc. becomes compromised.
		Delegitimization of hierarchy-legitimating myths	Challenges to ideologies, narratives and policies that validate hierarchical power arrangements.
		Structural victimization of LPGs	Denial of identity, security, and voice
		Structural violence toward LPGs	Unequal access to housing, health care, nutrition, education, etc.
		Lack of awareness	An insulated and inattentive HPG
		Accumulation of indignities	Pervasive patterns of "civilized oppression" by HPGs against LPGs
	Instability and anarchy	Seismic shifts	Periods of rapid social change and instability
		Tainted infrastructure	Compromised institutions, laws, and social norms for conflict regulation
		Looking up	Changes in LPGs' aspirations
		Power shifts	Changes in the balance of power between HPGs and LPGs
		Ambiguity of power	Unclear relative status of groups in conflict leading to more volatility
		Anarchy	The complete collapse of social order.
Issues	Nature of core issues	Dialogic poles	Underlying issues rife with apparent trade-offs
		Paradoxical dilemmas	Issues that when resolved create new problems
		Intricate interconnections of issues	Complex connections between distinct issues
		High centrality	Issues that have high personal or group-based importance
	Meaning	Truth	Issues that revolve around important, basic beliefs
		Hub issues	Grievances embedded within broad beliefs, ideologies, and basic assumptions

Relationships	Nature of relationships	Exclusive structures	Structures that keep groups isolated and out of contact with each other
		Inescapable relationships	Relationships from which it is virtually impossible to exit
		Collapsed relationships	Relationships damaged beyond repair by conflict
		Intense mixed motives	High-stakes conflicts with a mix of cooperative and competitive goals
		Intractable core	Fundamentally unsolvable issues
	Identity	Polarized collective identities	Group identities based on the negation and destruction of the "other"
		Conflict identities	Group identities organized around an ongoing conflict
		Monolithic and exclusive identities	All different aspects of ingroups and outgroups collapse into single entities
		Frozen identities	Personal and group identities become rigid and unresponsive to change
	Internal dynamics	Unconscious needs and defenses	Motives that are operative but difficult to identify and address
		Intragroup divisions and factions	When internal group divisions drive intergroup conflict
		Hidden agendas	Covert or criminal objectives that drive the overt conflict
Processes	Emotions	Emotional contagion	The pervasive spread of toxic emotions such as humiliation, deprivation, loss, rage
		Memorialized conflict	Conflict driven by a sense of duty and loyalty to those lost in war
		Socially-constructed volatility	Ingroup processes which create rules and norms that sanction destructive emotions
		Impaired cognitive functioning	Information processing impairment resulting from protracted, high-intensity situations.
	Social processes	Malignant social processes	Self-perpetuating, inescapable emotional dynamics
		Escalatory spirals	Basic tit-for-tat escalatory dynamics that run amok
		Structural changes	Changes in social and institutional structures due to escalation which perpetuate conflict
		Moral exclusion	Conditions where groups see out-groups as deserving of immoral treatment
		Violent exchanges and atrocities	A tipping point when violence justifies and begets more violence
	Pervasiveness and complexity	Pervasiveness	Conflict spreads into functional aspects of life (education, cultural systems, etc.) and transforms them into tools of conflict
		High complexity	When conflicts become too complex to comprehend
		Multilevel	When negative aspects of conflicts link from people to groups to institutions to cultures

(continued)

Table 2.1 (continued)

		Multiparty	Increasing numbers of stakeholders contribute to its perpetuation
		Chaotic and mercurial	Constantly changing dynamics which perpetuate conflict
Outcomes	Trauma	Individual and community trauma	Families and communities lose the capacity to trust one another and therefore function
		Betrayal of trust	A rupture of the basic understanding of a predictable world
		Beyond posttraumatic stress disorder	When atrocities lead to extraordinary levels of trauma that impair basic functioning
	Duration	Trauma unaddressed	When past trauma is left untreated to fester
		Historical rivalries	Robust, long-term animosities among people and groups
		Enduring cycles of low to high intensity	Cycles of conflict that alternate between high and low intensity, which can lead to temporary states of complacency.
		Destructive norms	Hostility and violence become the accepted norm
		Intergenerational perpetuation	Children and newcomers are socialized into the conflict
		Lasting commitments	A conflict's long duration justifies its perpetuation

between a powerful majority group and members of a low-power minority group in East Asia. Such disputes may differ in terms of the number of parties and issues involved, the relative ease of separation of the parties and enforceability of agreements, and the degree of involvement of outside representatives, institutions and bureaucracies (see Rubin & Levinger, 1995, for an excellent discussion of cross-levels comparisons).

However, might the diverse, paradoxical nature of intractable conflict actually belie something more fundamental that could allow for integrative understanding? Is conflict, at base, paradoxical or can it be conceptualized in a way that is consistent with general models of human behavior? Can theoretical coherence be achieved for a fixture of human relations that spans levels of social reality, from dyads to international relations? These are the questions that peace and conflict scholars have wrestled with for almost a century.

Five Paradigms of Conflict Studies

Over the past several decades, the literature on peace and conflict has generated a large array of approaches for the study of protracted social conflict (see Coleman, 2004). These perspectives have emerged from a variety of disciplines such as political science, social psychology, developmental psychology, law, education, communications, anthropology, linguistics, public health, and economics. Below, we outline five major paradigms currently employed in framing research and practice in this area: *realism*, *human relations*, *pathology*, *post-modernism*, and *systems*. These paradigms are, in effect, clusters of approaches that vary internally across a myriad of important dimensions, and overlap to some degree with approaches from other paradigms. The five paradigms are presented in order from most to least influential in the field today.

The Realist Paradigm

Historically, this perspective has been the dominant paradigm for the study of war and peace in history, politics, and international affairs. Essentially a political metaphor, it views protracted conflicts as dangerous, high-stakes games that are won through strategies of domination, control, and counter-control (see Schelling, 1960). Although they vary, approaches of this nature tend to assume that resources and power are always scarce, that human beings are always capable of producing evil and have a will to dominate, and that one's opponents in conflict at any point may become aggressive. Consequently, they present an inherently conflictual world with uncertainties regarding the present and future intentions of one's adversary leading to risk-aversive decision-making. Thus, intractable conflicts are thought to result from rational, strategic choices made under the conditions of the "real politics" of hatred, manipulation, dominance, and violence in the world. These conflicts are seen as "real conflicts" of interest and power that exist objectively due to scarcities in the world, and are only exacerbated by such psychological phenomenon as fear, mistrust, and misperception. In this context, power is seen as both paramount and corrupting, and real change is believed to be brought about primarily through power-coercive, command and control strategies. Neorealism, a less ardent version that emphasizes the constraints imposed by international structures, operates with many of the same assumptions (Waltz, 1979).

At the same time, though, this orientation to conflict is based on assumptions of rational choice that are "economic" in nature (i.e., reasoning through efficient cost/ benefit analyses), which, although valid under certain conditions, fail to account for many other types of human reasoning and action (such as social, legal, moral, and political forms of reasoning), which function differently and have a large impact on decisions and outcomes in conflict settings (see Diesing, 1962, for an extensive discussion). In addition, its "preventative orientation" to managing conflict (see Higgins, 1997) leads to a focus on short-term security needs, worse-case scenarios, and an over-reliance on strategies of threat and coercion (see Levy, 1996). Furthermore, its core competitive assumptions (regarding the nature of power and security, the availability of resources, and the inevitability of the other's aggression) can limit a party's response options and typically results in competitive and escalatory dynamics and self-fulfilling prophesies which foster further entrenchment in the conflict (see Deutsch, 1973, 2000).

The Human Relations Paradigm

An alternative to the realist paradigm emerged primarily through the social-psychological study of conflict (discussed below), and stresses the vital role that human social interactions play in triggering, perpetuating, and resolving conflict. Based on a social metaphor, its most basic image of intractable conflict is of destructive relationships in which parties are locked in an increasingly hostile and vicious escalatory spiral, and from which there appears to be no escape. With some variation, these approaches view human nature as mixed, with people having essentially equal capacities for good and evil, and stress the importance of different external conditions for eliciting either altruism and cooperation or aggression and violence. This orientation also identifies fear, distrust, misunderstanding, and hostile interactions between disputants and between their respective communities as primary obstacles to constructive engagement. Thus, subjective psychological processes shape disputants' perceptions, expectations, and behavioral responses, thereby playing a large role in determining the course of conflict (see Deutsch, 1973). From this perspective, change is thought to be brought about most effectively through the planful targeting of people, communities, and social conditions, and is best mobilized through normative-reeducative processes of influence (Fisher, 1994).

Nevertheless, relationally-focused strategies of intervention, when not complemented by other methods, often fall well short of their objectives in hazardous situations of protracted conflict. Although overstated, they have been criticized by some realists as "...at best well-intentioned, at worst soft and driven by sentimentalism, and for the most part irrelevant" (Lederach, 1997). They typically work best in situations where there is an a priori acceptance of the values of reciprocity, human equality, shared community, fallibility, and non-violence (Deutsch, 2000). Contexts that are void of these norms, and of the laws and institutions that regulate them, present substantial challenges to the constructive use of relational strategies. For example, in societies where male-superiority goes unquestioned, the use of cooperative strategies to address protracted gender conflicts may in fact only perpetuate the oppressive quality of gender relations in that context. Finally, most human relations approaches are based on the values and assumptions of scientific humanism and planned social change (Fisher, 1994). These values and assumptions define the boundaries of these approaches and limit their applicability in situations where such values are not shared.

The Medical Paradigm

This paradigm pictures intractable social conflicts as pathological diseases—infections or cancers of the body politic—that can spread and afflict the system and that need to be correctly diagnosed, treated, and contained. In this view, the conflict system (the patient) is a complicated system made up of various interrelated parts that can be analyzed and understood directly and treated accordingly. Conflicts are thought to be treated most effectively by outside experts who have the knowledge, training, and distance from the patient necessary to accurately diagnose and address the problem. This perspective views humans and social systems as basically health-oriented entities that, due to certain predispositions, neglect, or exposure to toxins in the environment, can develop pathological illnesses or tendencies that are destructive. Treatment of these pathologies, particularly when severe, is seen as both an art and a science, with many courses of treatment bringing their own negative consequences to the system. Although not as common as the realist and human relations paradigms, the medical model is particularly popular with agencies, community-based organizations, and non-governmental organizations working in settings of protracted conflict.

However, once again, this worldview is limited in its capacity to manage protracted conflict unaided. The practice of medicine often seeks to address symptoms of illness (coughs, fever, rashes, etc.) rather than underlying causes, just as we often seek to address the expression of a pathology with Band-Aid solutions (e.g. armed violence treated by ceasefires) rather than addressing the systemic problem that gave rise to the emergent symptom. Although hostilities between people may be temporarily controlled by the acceptance of a cease-fire or peacekeeping troops, the conflict may move no closer to resolution and may in fact become more intractable as a result of the disengagement of the parties (Fisher, 1997). In addition, the approach of identifying and exposing covert or unconscious motives and interests rests on the assumption that doing so is good. Thus, it assumes that it is both possible and constructive to unearth such motives, that people have the capacity and support to tolerate such information when it is forthcoming (about themselves, their government, their businesses, and so on), and that people, corporations, and governments will then have the motivation and the capacity to reform. These assumptions, although hopeful, are often inaccurate. Finally, this orientation is based on a deficit model, with a focus on that which is wrong or pathological in a conflict system. Although important, this orientation often neglects focusing on positive responses such as resiliency or altruistic and ethical behavior under difficult circumstances, and can foster a negativity bias in our understanding of and responses to the phenomena.

The Post-Modern Paradigm

Also known as constructivism, this perspective portrays intractable conflicts as rooted in the ways we make sense of the world. A linguistic and communications metaphor, its most basic image is of conflict as a story: a narrative or myth that provides a context for interpretation of actions and events, both past and present, that shapes our experience of ongoing conflicts. Thus, conflict springs from the way parties subjectively define a situation and interact with one another to construct a sense of meaning, responsibility, and value in that setting. Intractable conflicts, then, are less the result of scarce resources, incendiary actions of parties, or struggles for limited positions of power, than they are a sense of reality, created and maintained through a long-term process of meaning-making through social interaction (Lederach, 1997; Pearce & Littlejohn, 1997).

This worldview highlights a form of power as meaning-control: an insidious, although primary form of power that is often quietly embedded in the assumptions and beliefs that disputing parties take for granted. It suggests that it is primarily through assumptions about what is unquestionably "right" in a given context that different groups develop and maintain incommensurate worldviews and conflicts persist. Thus, change is believed to be brought about by dragging these assumptions into the light of day through critical reflection, dialogue, and direct confrontation, consequently increasing disputants' awareness of the complexity of reality, of our almost arbitrary understanding of it, and of the need for change.

Although rich and intuitively appealing, post-modern constructivism has been criticized for its abstract intellectualism (Alvesson & Willmott, 1992a, 1992b) and its tendency to denigrate and alienate the elite (Voronov & Coleman, 2003). Critics find its central ideas and jargon vague and difficult to operationalize in any useful manner. It seems to find meaning-making processes and dominance everywhere, but makes it difficult to pinpoint them anywhere. It has also been chided for its over-emphasis on the subjective and concomitant denial of the importance of objective circumstances. Although intriguing, this level of consciousness can be quite demanding and difficult to sustain, even under non-threatening conditions (Kegan, 1994). Therefore, the possibilities of applying such methods in situations of intense, protracted conflict are especially challenging.

The Systems Paradigm

In essence, the system's perspective is based on an image of a simple living cell developing and surviving within its natural environment. A biological metaphor, it views conflicts as living entities made up of a variety of interdependent and interactive elements, nested within other, increasingly complex entities. Thus, a marital conflict is nested within a family, a community, a region, a culture, etc. The elements of systems are not related to one another in a linear manner, but interact according to a non-linear, recursive process so that each element influences the others. In other words, a change in any one element in a system does not necessarily constitute a proportional change in others; such changes cannot be separated from the values of the various other elements which constitute the system. Thus, intractable conflicts are viewed as destructive patterns of social systems, which are the result of a multitude of different hostile elements interacting at different levels over time, culminating in an ongoing state of intractability (see Coleman, Vallacher, Nowak, & Bui-Wrzosinska, 2006; Pruitt & Olczak, 1995). Power and influence in these systems are multiply determined, and substantial change is thought to occur only through transformative shifts in the deep structure or pattern of organization of the system.

Important insights for working with intractable conflict follow from three basic assumptions of general systems theory: the non-linear nature of the relations between elements in a system; systemic openness and the importance of the internal and external environment; and the structure of nested systems within systems. These ideas shift our thinking away from simple, essentialized, static, or dyadic views of conflict. In its place, we can begin to understand the complex, multi-level, dynamic, and cyclical nature of these phenomena. However, general systems theory has been criticized for its lack of specificity, for its imprecise definition, and for contributing relatively little to the generation of testable hypotheses in the social sciences (Kozlowski & Klein, 2000). Work from this perspective will need to move beyond its use as a general heuristic in order for it to realize its full potential to addressing complex social conflicts. This has been a primary goal of the development of dynamical systems' principles and methods for applications to conflict and peace dynamics.

The five paradigms outlined in this section provide us with an extensive menu of perspectives and options for viewing intractable social conflicts. Each approach is supported to some degree by empirical research, and each offers a unique system of questioning that governs the way we think about intervention in conflicts. Ideally, however, we must develop a capacity to conceptualize and address intractable conflicts that is mindful of the many factors and complex relationships inherent in such phenomena, and of the complementarities of these diverse approaches.

The Social Psychology of Conflict

Social psychology was born in the cradle of war (Morawski, 2000) and has been compelled by world events to study conflict throughout its history. Ablion Small, a sociologist, suggests that social psychology arose from a time when Americans "whose thought-world had been stirred to its depths by the war found themselves in 1895, star-gazing in social heavens that had never looked so confused nor so mysterious" (quoted in Karpf, 1932, p. 213). Others have identified World War I and II as two of the major forces behind the early development of social psychology (Allport, 1954; Cartwright, 1979; Jones, 1985). Scholars have even gone as far as crediting Adolph Hitler and the Nazi Party in Germany with igniting social psychology, as the fear and oppression imposed by the Nazi's forced pioneers of the field like Kurt Lewin to flee Europe and to come to the U.S. to study problems such as conflict, prejudice, and authoritarianism (Cartwright, 1979; Jones, 1985). These beginnings have carried the field far in its understanding of conflict, violence, and war, largely through the work of prominent psychologists such as Lewin, Jerome Frank, Charles Osgood, Morton Deutsch, Harold Kelley, Herbert Kelman, Dean Pruitt, Dan Druckman, and David Johnson and many others.

Classic Perspectives

Several classic lines of theory and research inform the social-psychological study of conflict today. Early theories of conflict presented it in the context of competitive struggles and employed a deterministic mode of explaining conflict, war, and hostility. This included Darwin's (1859, 1871) evolutionary perspective of the struggle for survival of the fittest in species, Freud's (1936, 1960) psychodynamic perspective of internal psychic struggles between the Id, Ego and Superego or between the conflicting internal drives of Eros (life, love, creativity, etc.) and Thanatos (death, aggression, destruction, and so on), and Marx's (1844/1997) socio-political-economic perspective of class struggle. Although these early works provide evidence for the claim that systemic and dynamic perspectives are at the very roots of the social-psychological approach to conflict studies, the lack and paucity of systematic empirical support for these perspectives rendered them more doctrines and ideologies than scientific theories. This was also largely true of many of the early "armchair-speculation" social-psychological theories of pleasure-pain, egoism, sympathy, gregariousness, imitation, and suggestion (Allport, 1969).

However, the subsequent transition of social-psychological research into a phase of more sophisticated, rigorous empirical science came largely at the cost of systemic, dynamic thinking. What the development of reductionist, systematic, and particularly laboratory methodologies delivered in terms of precision and predictive power of research, they lost in terms of the more complex, holistic, temporal conceptualization of conflict dynamics. Ultimately, this led to both the specialization and the fragmentation of the field at its early stages of conflict theory development, where the socio-political, economic, psychological, and evolutionary perspectives became largely disconnected.

This basic disconnect is still visible today in the lack of a common scientific platform for uniting the main areas of conflict studies, such as political and economic science, psychology and social psychology, law, evolutionary biology and primatology, anthropology, and epidemiology. Typically, psychological models privilege cognitive factors as central mechanisms in social, economic, and political conflict processes, while socio-economic models emphasize structural, social and economic factors in determining human functioning in conflict. This results in an understanding of the effects of economic, social, and political structures that is disconnected from our sense of human factors and dynamics.

Two basic challenges to the coherence of the social-psychological approach to conflict come from the considerable pressure in empirical research to identify (linear) causal mechanisms, and the dearth of multi-level models and methodologies for conducting research. However, the genesis of social-psychological thinking was built on the assumption that social-structural and psychological conditions and processes are intimately linked. Aristotle and Plato, arguably two of the earliest social-psychological thinkers, spent considerable time contemplating and debating the interactive relationship between the individual and society (Allport, 1969).

Later, Kurt Lewin offered his elegant formula for conceptualizing the interaction between individuals (P) and the environment (E) in determining human behavior (B): B = f (P x E). This framework encouraged unity in a young discipline that was increasingly divided between the social and personality schools, and offered a whole new vocabulary for the conceptualization and scientific study of conflict processes (discussed in greater detail in Chap. 3). This included such constructs as force fields, tension systems, vectors, temporal trajectories, goal gradients, systemic and reciprocal causation, as well as the importance of understanding the complexity of human motivation and the value of mathematization of psychological constructs. These and other contributions of Lewin (e.g., action research, field theory) highlighted the circular causation between the individual and the environment (see Deutsch, 1969), a tradition that finds a continuation in the dynamical systems approach to conflict today.

Social-Psychological Models of Social Conflict

Six theoretical models of social conflict build on elements of the Lewinian approach have been particularly influential in the development of the field: *game theory*, *social interdependence theory*, *social motivation theory*, *dual-concern theory*, *power dependence theory* and *social identity theory* (see Coleman et al., 2012). Each of these models has focused generally on *understanding the conditions and processes that lead to constructive versus destructive conflict dynamics and outcomes* (see De Dreu & Gelfand, 2008; Deutsch, 1973, 2002; Deutsch, Coleman, & Marcus, 2006). They are summarized below.

Game Theory

The Cold War era (1950–1980s) led to a stronger emphasis in social psychology on conflict structure (realpolitik) and the dominance of the game theoretical perspective in conflict conceptualization, methodology, and policy-making (Schelling, 1960; Von Neumann & Morgenstern, 1944). In 1944, Von Neumann and Morgenstern published their now famous *Theory of Games and Economic Behavior*, which formulated conflicts of interest in mathematical terms. Like Lewin's (1947, 1948) work in psychology, this approach emphasized the interdependent nature of disputants' interests, behavior, and fates. However, it focuses on rational decision-making in conflict, positing that there is always a "rational" choice that is the best counter-choice to that of one's opponents.

This approach has been most effective in predicting behaviors and outcomes in "zero-sum" situations, where the maximization of one party's outcomes means the minimization of the other party's outcomes. Scholars contend that the game theoretical approach is most predictive in competitive situations because of the underlying assumptions inherent in the theory of what constitutes "rational" choice (e.g., efficiency, maximization of goals, transitivity of preferences, etc.; Deutsch, 1985; Diesing, 1962). These assumptions limit the applicability of the model to more distributive, economic types of conflicts.

Conflict research from a game theoretical perspective has focused largely on the rules and strategies for *winning* conflict games,[1] but it also investigates the structure and process of interdependence between parties, with a focus on identifying the conditions for achieving a state of equilibrium or stability between them. Particularly important is the recognition that many if not most conflict situations are of a mixed-motive (cooperative and competitive) nature (Schelling, 1960). An important finding from this research is that players pursuing only their self-interest in

[1] Central to this was the development of the "mini-max" strategy, where negotiators strive to develop strategies that limit the alternatives available to their adversaries so that when their adversaries choose to minimize their maximum loss, they will inevitably choose an alternative that is good for their side.

these games tend to end up worse-off over time than players who consider the other's concerns when making choices. The prisoner's dilemma (e.g. Axelrod, 1984), for example, has been used to investigate the strategies individuals employ under incentive structures that pit self-interest against cooperation and collective interest. This approach has shown that although competition is clearly not the best strategy, neither is consistent cooperation because it sets up the individual for exploitation. The best strategy for maximizing individuals' outcomes is responsiveness to one another's behavior in a "tit-for-tat" manner, with a readiness to move to cooperation in line with a "forgiveness" mindset.

Methodologically, game theory has had a huge impact on conflict research as it introduced game matrices, which are abstract representations of social conflicts that are precise and efficient and allow for the investigation of dynamics *over time*. In 1977, Pruitt and Kimmel documented that over 1,000 studies had been published employing experimental games. However, as Deutsch writes, "Much of this research...was mindless—being done because a convenient experimental format was readily available" (Deutsch, 2002, p. 313). In other words, although the research shed light on more rational, competitive conflict dynamics, no grand theory to better inform our understanding of conflict dynamics more generally emerged from game theory (Pruitt, 1998). Nevertheless, conflict scholars continue to work with these ideas and methods today, and it can be said that the mathematical approach of dynamical systems to conflict has been partially in response to the values and limitations of past game theoretical approaches.

Social Interdependence Theory

Based on Lewin's insight that interdependence was the essence of group dynamics (Lewin, 1936, 1948), this theory specified the basic conditions and processes involved in constructive versus destructive conflict (Deutsch, 1973). Deutsch's earlier research (Deutsch, 1949a, 1949b) showed how different *types of goal interdependence* between parties—*positive* or cooperatively linked goals versus *negative* or competitively linked goals—affect constructive versus destructive processes and outcomes in groups, respectively. When applied to conflict, the research found that constructive processes of conflict resolution are similar to cooperative problem-solving processes where the conflict is seen as a mutual problem by the parties and addressed jointly; and that destructive conflict processes are similar to competitive processes where the conflict is framed and approached as a win-lose struggle (Deutsch, 1973).

This basic idea cascaded into a variety of propositions (outlined in Deutsch, 1973) that provide a general intellectual framework for understanding conflict and how to intervene constructively. Deutsch's theory of conflict resolution is widely recognized as one of the most important advances of the twentieth century in the study of conflict (Jones, 1998). It has been validated by a large canon of empirical studies (see Deutsch, 1973; Johnson & Johnson, 1989, 2005), and has led to a wide array of practical methodologies and training strategies for the constructive resolution of conflict (see Coleman & Deutsch, 2001; Coleman & Lim, 2001; Deutsch et al., 2006; Johnson & Johnson, 2005; Lewicki, Saunders, Barry, & Minton, 2004; Tjosvold, 1991; Tjosvold & Johnson, 1983).

However, Deutsch's theory has its limitations (see Deutsch, in press; Johnson & Johnson, 2005). Two of the most consequential are its assumptions regarding social power and interdependence. The original formulation of the theory assumed both *equal power* and *high degrees of interdependence* between the parties in conflict (Johnson & Johnson, 2005). Thus, the processes and outcomes observed in most of the empirical studies supporting the original theory occurred under these conditions.

Social Motivation Theory

Another influential model of social conflict, informed by the work of Kelley and Thibaut (Kelley & Thibaut, 1978; Thibaut & Kelley, 1959) on interdependence and McClintock (1976) on social motives, describes how both individual and situational differences in interdependence affect people's *social orientations* and thus their values and behaviors when negotiating disputes (see De Dreu, Beersma, Steinel, & Van Kleef, 2007; Van Lange, De Cremer, Van Dijk, & Van Vugt, 2007). Although a variety of different social motives have been identified (e.g., altruistic, competitive, and individualistic), most of the research on negotiation and conflict has focused primarily on the effects of *pro-self* versus *pro-social* motives. Pro-self motivation combines both individualistic and competitive goals into one motivational orientation, whereas pro-social motivation combines both cooperative and altruistic goals.

Research has shown that negotiators with a pro-self motive seek to maximize their own outcomes and have little or no regard for the other party's outcomes, tend to see negotiation as a competitive game where wielding power and winning are key, and selectively search and process information that is consistent with this competitive view (De Dreu et al., 2007). In contrast, negotiators with a pro-social motive seek fair outcomes that maximize both one's own and the other party's goals, tend to view negotiations in more cooperative terms where harmony, solidarity and fairness are important, and seek information that validates this view. This research has found that while social motives have a strong dispositional component (De Dreu & Van Lange, 1995), the tendency to adopt a pro-self or pro-social orientation can also be triggered by reward structures (e.g., Weingart, Bennet, & Brett, 1993) and social cues (e.g. Burnham, McCabe, & Smith, 2000).

The research on social motives has highlighted the importance of motivational orientations in affecting short-term responses to conflict, and sheds light on some of the social-cognitive processes associated with pro-self and pro-social motives. But it has two significant limitations. First, by collapsing together distinct motives such as competition and individualism in conflict, it often *conflates* what may in fact be important conflicting tendencies in some social situations (maximizing one's own outcomes versus accepting less in order to defeat the other party). Second, research on social motives has tended to investigate the comparative effects of distinct motivational orientations (cooperation, egalitarianism, altruism, etc.) in conflict, but has yet to fully address how these basic aspects of social relations can be *integrated* into an account of how social motives function together in conflict (see Van Lange et al., 2007).

Dual-Concern Theory

This approach was originally conceived as a model of individual differences in conflict resolution styles (Blake, Mouton, Barnes, & Greiner, 1964; Filley, 1975; Rahim, 1983, 1986; Thomas, 1976), but has subsequently been developed into a more predictive theory of choice and strategy under different motivational conditions in conflict (Pruitt & Kim, 2004; Pruitt & Rubin, 1986). It proposes that differences in two basic concerns of disputants—concerns for their own outcomes and concerns for the other parties' outcomes (ranging from weak to strong)—combine to affect the strategies people choose in conflict, including *yielding*, *avoiding*, *contending*, *problem-solving*, and *compromising*. With this model, Thomas (1976) extended Deutsch's one dimensional model (competitive or cooperative) to include a second dimension, as he saw self-concern and other-concern as orthogonal interests, which can function independently. These two concerns can differ according to social conditions (reward structures, social and cultural norms, etc.) and individual differences in style preferences.

Dual-concern theory has received some empirical support (e.g., Sternberg & Dobson, 1987; Van de Vliert & Kabanoff, 1990) and has begun to address how motivational differences (concerning both self and other) can operate orthogonally and vary by *degrees of importance*. Subsequent research on this model has primarily focused on the conditions that foster self and other concerns in conflict, but has also indentified how strategic choice is moderated by peoples' resistance to yielding (Druckman, 1994; Pruitt, 1998) and the perceived feasibility of employing different strategies in particular contexts (Kelley, 1967; Pruitt, 1981; Pruitt & Kim, 2004). However, the model has yet to address how differences in the *power* distribution of parties affect people's strategies and choices in conflict situations. This shortcoming limits the generality and predictive power of the model.

Power Dependence Theory

Many studies of power differences in negotiations are based on the theory of power dependence (e.g., Gerhart & Rynes, 1991; Kim & Fragale, 2005; Mannix, Thompson, & Bazerman, 1989; Ng, 1980). Situated within the broader framework of interdependence theory (Kelley & Thibaut, 1978; Thibaut & Kelley, 1959), power dependence theory states that "the power of A over B is equal to and based upon the dependence of B on A" (Emerson, 1962, pp. 32–33). Dependence is based on two dimensions of the negotiation situation (1) it is directly proportional to the value attributed by a party to the outcome at stake, and (2) it is inversely proportional to the availability of this outcome through alternative sources. Laboratory research on negotiations has generally supported this model, finding that negotiators who hold more attractive BATNAs (Best Alternatives to a Negotiated Agreement—or the possibility of achieving desired outcomes through alternative means) or who are able to increase the other party's dependence, are less dependent on their negotiation partners and thus possess greater power relative to them and obtain better outcomes in negotiations (Kim, 1997; Kim & Fragale, 2005; Mannix, 1993; Pinkley, Neale, & Bennett, 1994).

Power dependence theory has been particularly predictive in the realm of distributive or competitive negotiations, but also has its limitations (see Kim & Fragale, 2005). Most notably, by defining and operationalizing power *solely* as "asymmetries of dependence", the model overlooks the many other types of power and influence operating in social relations to shape conflict dynamics. Beyond that, it does not incorporate the potential for *change* in the differential dependencies among parties to a conflict over time (Kim et al., 2005).

Social Identity Theory

Some conflicts become deeply anchored in the way people define themselves and their own groups, which contributes greatly to their intractability. Social Identity Theory (SIT; Tajfel & Turner, 1979) has advanced the understanding of the individual psychological processes at the root of ethnocentrism and intergroup conflict by emphasizing the link between group identification and intergroup relations. SIT was developed in the context of research on realistic conflict theory (Campbell, 1965; Sherif, Harvey, White, Hood, & Sherif, 1961), where intergroup hostility is viewed as the result of competition over real or perceived scarce resources. Researchers set out to identify the minimum baseline for resource scarcity that would trigger intergroup conflict, and were surprised to find that simply categorizing people into arbitrary groups seemed to elicit ethnocentric ingroup bias (Tajfel, 1970). These findings seemed to indicate that scarcity of resources was not a necessary precondition for intergroup conflict, since the introduction of even minimal group distinctions was sufficient for initiating competitive intergroup relations.

Building on the results from the minimal group paradigm, Tajfel and Turner (1979) developed a model proposing a link between group categorization and tendencies toward ingroup bias and out-group discrimination. Essentially, social identity theory suggests that people are motivated to achieve and maintain positive self-esteem, and one avenue for achieving this is through positive ingroup associations when compared to outgroups. In this perspective, out-group derogation provides a means of boosting one's self-esteem as a member of the ingroup. In extreme cases, the mere presence of an out-group may be seen as a threat to a positive social identity or even to a group's existence (Kelman, 1999), and thus fuel conflict intractability due to a zero-sum perception of the opposing groups identities. Research on social identity theory has focused recently on identity content (Hewstone & Cairns, 2001; Lalonde, 2002; Turner, 1999; Turner, Oakes, Haslam, & McGarty, 1994) and categorization processes (Perdue, Dovidio, Gurtman, & Tyler, 1990), and has been applied to cases of intractable conflict such as Northern Ireland (Cairns & Darby, 1998; Livingstone & Haslam, 2008). It should be noted, though, that the link between self-esteem and out-group derogation is controversial. Research on collective self-esteem (e.g., Crocker & Luhtanen, 1990; Luhtanen & Crocker, 1992) and collective narcissism (Golec De Zavala, Cichocka, Eidelson, & Jayawickreme, 2009) has established an intricate link between self-esteem and intergroup conflict at the individual and group level.

Although social identity theory has contributed to better understanding of collective behavior including ingroup bias, responses of subordinate groups to their positions of unequal status, and intragroup homogeneity and stereotyping, the relationship between group identification and ingroup bias is still unclear. In addition, the generalizability of the self-esteem hypothesis across different cultures, as well as its relevance for asymmetric conflicts and low-status groups, has been questioned (Brown, 2000), and it has also been criticized for its lack of predictive power (Hogg & Williams, 2000; Turner & Reynolds, 2001). In one meta-analysis, Hinkle and Brown (1990) found that across 14 studies, the overall association between identification and intergroup bias was negligible. Although social identity theory may explain some key mechanisms at play in intergroup conflict, the multitude of other factors contributing to conflict and intractable conflict in particular (see Coleman, 2003) may limit the role of identity in fully explaining the dynamics of intractable conflict.

The six models outlined in this section have made considerable inroads into our understanding of constructive versus destructive conflict in social relations. However, because these inroads have often proceeded in somewhat different directions, we today find ourselves with a rather fractured understanding of conflict dynamics. We currently have a series of mid-level or micro-level models of conflict antecedents, processes, and outcomes that have yet to become convergent with a more general theory of social relations. Clearly, the time is ripe for a more comprehensive approach to social conflict.

Contemporary Social Psychological Themes

A concern with the nature of social conflict is also represented in several contemporary approaches to conflict. These approaches span different levels of analysis, incorporate different aspects of conflict processes, and consider different characteristics of conflict and of the conflicting parties. Below we provide a brief overview of these orientations by focusing on the particular themes they address.

Can Conflicts Have Positive Consequences?

During the last 15 years, there has been an interest in exploring the potentially positive consequences of conflict. The basic idea is that dissent based on differences in knowledge, opinions, and expertise can be beneficial when such views are integrated constructively (De Dreu & Weingart, 2003). In this context, task conflict—which refers to conflicts about the distribution of resources, procedures and policies, judgments and interpretations—has been found to enhance understanding of the matter at hand and thus lead to better performance as well as increased creativity and innovation (Schulz-Hardt, Mojzisch, & Vogelgesang, 2008). Even though meta-analyses (De Dreu & Weingart, 2003; De Wit, Greer, & Jehn, 2012) do not support the positive consequences of task conflict per se, it is to date assumed that under specific circumstances positive consequences of conflict can be found. For example, the motivated information processing perspective assumes that cooperative outcome interdependence might set the stage for constructive and task-relevant exchanges of information, but only when individuals engage in systematic information processing, in order to combine and integrate the information held by individuals (e.g., Bechtoldt, De Dreu, Nijstad, & Choi, 2010; Brodbeck, Kugler, Fischer, Heinze, Fischer, 2011; De Dreu, Beersma, Stroebe, & Euwema, 2006). In contrast, relationship conflict—which refers to conflicts about personal taste, political preferences, values, and interpersonal style—appears to lack the potential for positive consequences.

Which Motives Drive Individuals' Behaviors in Conflict Situations?

In most conflicts, goal interdependence is neither purely cooperative nor purely competitive but rather reflects mixed motives (Katz & Kahn, 1978). Accordingly, contemporary research has focused on how bundles of motives drive individuals in such situations. Mixed-motive situations offer incentives to invest in the collective goal as well as in personal benefit. Depending on personal factors as well as situational influences, mixed-motive conflict scenarios promote either pro-social motives or pro-self motives. Whereas pro-social motives drive individuals to focus on a common benefits and goals, pro-self motives are linked to goals and concerns with personal benefit (De Dreu, Weingart, & Kwon, 2000).

What Factors Activate Biases and Stereotypes Regarding Out-groups in Conflict Scenarios?

It is not surprising that extremists openly favor their own in-group *vis a vis* out-groups, and that they harbor prejudice and stereotypical views regarding out-groups as well (Fiske, 2002). These features of in-group—out-group relations, however, also characterize people who are moderate in their views and not in direct conflict with an out-group. These biases are mostly automatic and unconscious (Dovidio & Gaertner, 1986; Fiske, 1998, 2000), but nevertheless can influence everyday behavior. For example, people tend to behave and speak in a less friendly manner and to withhold positive behaviors that suggest liking and respect toward out-group members. Interestingly, there is evidence that out-group cues heighten activity in the amygdala, the center in the brain of fear and anxiety (Fiske, 2002). This is true for moderates as well for those with explicitly antagonistic views of out-group members. There is debate, however, whether these biases are unconditionally versus conditionally automatic (i.e., unlearned vs. learned), since factors such as education, economic opportunity, and constructive contact among groups have been found to decrease the biases and their consequences.

What Promotes and Maintains Oppression Between Groups?

Social dominance theory (Sidanius, Pratto, van Laar, & Levin, 2004) focuses on individual and structural factors that contribute to various forms of oppression (e.g., discrimination, racism, ethnocentrism, classism, sexism) among social groups. Oppression itself is seen as a specific form of a more general human tendency to form and maintain group-based hierarchies. Sidanius, Levin, Federico, and Pratto (2001), for example, argue that humans have a desire for unequal relations among social groups, regardless of whether this means in-group domination or in-group subordination. In these unequal relations, oppression is driven by systematic institutional and individual forms of discrimination, evident in the disproportional allocation of desired goods (e.g., prestige, wealth, power, food, health care) to the privileged groups and the greater allocation of undesirable things (e.g., dangerous work, disdain, imprisonment, premature death) to members of less privileged groups. The research on social dominance theory studies the determinants of group-based oppression at multiple levels of analysis, ranging from the individual within a group to the entire system in which groups are embedded.

Why Do Conflicts Endure?

System justification theory explores majority and minority group investment in the status quo by focusing on the "process by which existing social arrangements are legitimized, even at the expense of personal and group interest" (Jost & Banaji,

1994, p. 2). The theory suggests that individuals are inclined to legitimize, justify, and maintain the status quo (Jost & Banaji, 1994), and it distinguishes three different motives or justification tendencies: *ego justification* (the need to develop and maintain a favorable self-image); *group justification* (the desire to develop and maintain favorable images of one's own group), and *system justification* (social and psychological needs to legitimize the status quo of the entire system). Members of advantaged groups—having a favorable self-image, group-image, and the need to maintain the status quo—do not have a desire to change a system. Members of the disadvantaged groups are only likely to engage in social change when ego justification and/or group justification motives are stronger than system justification needs and tendencies (Jost, Banaji, & Nosek, 2004).

What Factors Foster Escalation and De-escalation in Conflicts?

Researchers have analyzed past conflicts to identify the factors that inhibit versus escalate conflict escalation. The factors identified thus far include the expansion of issues, the polarization of relations, the involvement of other biased parties, dehumanization of the adversaries, and a homogenous in-group (Kriesberg, 2003). On the other hand, factors such as sympathy and empathy, new interests, shared norms, modeling de-escalating conflict strategies, a loss of faith in the justice or legitimacy of the goals as well as growing interdependence between the parties have been identified as mitigating escalatory trends (Kriesberg, 2003).

When Is the Optimal Moment for Attempts at Conflict Intervention or Resolution?

Research on ripeness theory (Zartman, 1989, 2000) has identified basic factors that determine when a conflict is ripe for an attempt at resolution. Zartman (2000) writes: "The concept of a ripe moment centers on the parties' perception of a mutually hurting stalemate, optimally associated with an impending, past, or recently avoided catastrophe. ... The other element necessary for a ripe moment is ... the perception of a way out, ... a sense that a negotiated solution is possible for the searching and that the other party shares that sense and the willingness to search too" (p. 228). However, ripeness theory has been criticized on many grounds including being too passive, static, and simplistic a model (Pruitt, 1997; Rubin, 1991), but has recently been reconceptualized in dynamical-systems' terms (Coleman, Hacking, Stover, Fisher-Yoshida, & Nowak, 2008).

The Promise of Dynamical Systems

Although classic and contemporary research on conflict has made great strides toward understanding the conditions and processes that determine whether conflict will take a constructive or destructive course, none of these ideas fully explain the paradoxical and diverse nature of intractable conflict. The ideas are especially wanting when it comes to the small proportion of conflicts that progress to intractability. This is where the core assumptions and insights of dynamical systems enter the picture. These assumptions and insights have transformed the way scientists conceptualize and investigate phenomena in the physical sciences and there is reason to believe that they will do the same for the social sciences in general, and for the study of social conflict in particular.

What new perspectives can dynamical systems give us in understanding the causes of conflicts and how to resolve them? Some of the important issues in conflicts are to understand (1) whether latent conflicts are triggered into active conflicts only when some variables exceed certain threshold values; (2) how the patterns of behavior in groups develop from the many and diverse interactions between individual people; (3) whether almost identical conditions do, or do not, result in similar conflicts; (4) whether seemingly random events are really due to undeterminable influences; and (5) how we can assess, and perhaps even predict, the statistical properties of the variables that describe a conflict.

Perhaps surprisingly, quite similar questions have been asked about the properties of physical and biological systems. The success of dynamical systems in answering such questions in those fields gives us the hope that these methods may also shed new light on social systems such as conflicts. We will now describe how dynamical systems have addressed similar questions in three different physical or biology systems: the motion of air in the atmosphere, the spread of infectious diseases, and the movement of sand in a sand pile. Our understanding of the use of dynamical systems in these physical and biological systems can then be our springboard in using dynamical systems to develop a new perspective in social psychology on conflicts.

Saltzman (1962) and Lorenz (1963) used dynamical systems to understand a highly simplified model of the motion of air rising and falling in the atmosphere over the earth. The air is heated by the surface of the earth and cooled by the top of the atmosphere. The solution of the mathematical equations that they formulated to represent the air shed light on issues similar to those listed above. First, when the temperature difference between the earth and the top of the atmosphere is small, heat flows from one patch of air to the next and all the air remains motionless. But, if the temperature difference is larger than a critical value, the air starts to move. This happens because the heat from the earth is enough to expand each patch of air reducing its density, making it float up in the atmosphere. How high it floats depends on how much it has expanded, how fast it moves, and the temperature in the atmosphere around it. Thus, when a certain variable (the temperature difference) exceeds a threshold, the entire nature of the system changes (issue #1). Second, the motion of individual patches of air organize into an overall global pattern (issue #2). The air circulates up on one side and then down on the other side of large long cylinders. This is an example of "self-organization" and "emergence" where the local interactions between the tiny parts of a complex system, here the small patches of air, produce a globally organized pattern.

Third, and perhaps the most unexpected, surprising, and deeply important result of the dynamical system analysis by Lorenz was that if the initial motion of the air was started at only very slightly different speeds, then the motions later could be remarkably different (issue #3). That is, almost identical initial conditions would lead to patches of air that were moving upward on one part of the earth in one case but downward at that same part of the earth in another case. This result became known as "sensitivity to initial conditions" or more commonly called the "Butterfly Effect", meaning that the small effect on the air of a butterfly beating its wings in Kuala Lumpur, Malaysia could determine whether or not, a week later, there is a violent thunderstorm over New York City.

This important finding on initial conditions led to a new way to interpret experimental data and to think about physical (or social) systems (Liebovitch & Scheurle, 2000). Because there will always be the tiniest difference in the initial conditions each time we run the same experiment no matter how closely we try to match them, the Butterfly Effect means that the time series of measurements can be different, even very different, when we repeat the same experiment. We expect though that there is something that should indeed be the same every time we run a similar experiment. In fact, Lorenz found that certain combinations of the variables that describe the motion of the air traced out a pattern that was the same each time. This pattern is called an *attractor* (issue #4). Even though the time series of the variables can be different every time, the attractor remains the same. This is because the combination of the variables of each time series traces out a slightly different part of the attractor, but it is still the same attractor. It is called an attractor because every time an experiment is run on the same system, the values of this combination of variables are always drawn to those found only on the attractor. The importance of attractors is that the variables of some dynamical systems, no matter their initial values, always evolve in time to only these certain values. We will see throughout this book the central role that attractors play in understanding social systems in general and intractable conflicts in particular.

Dynamical systems have also been used to shed light on the spread of infectious diseases such as chickenpox, measles, and mumps. Over the last 100 years (Anderson & May, 1991) SEIR mathematical models have been used to describe the time course of how the number of susceptible individuals (S) become exposed to disease (E), develop the infection (I), and then recover (R). These equations are deterministic, meaning that the number of people in each S, E, I, and R group at any one moment can be directly computed from the number at the previous moment. Even though that is the case, the number of people infected each winter can vary in a seemingly random way (Schwartz, 1992) (issue #5). That such deterministic systems can produce behavior that seems to be due to chance became known as "chaos". Contrary to its usual meaning, here chaos means that a deterministic system, which is completely uninfluenced by chance, can generate effects so complex and unpredictable that they appear to be due to chance. That seeming random social behaviors can be produced by deterministic social mechanisms is another useful lesson for social psychology from the study of dynamical systems in physics and biology.

Dynamical systems have also been used to gain a better understanding of how local interactions between small parts in a physical system can produce "emergent" global behavior. For example, in many physical systems, added stress is tolerated between small parts until a threshold is exceeded and then that stress is relieved by being distributed to the neighboring parts. The eponymous example is that of a sand pile. Sand is added at the top until the local slope exceeds a critical value sending some sand downhill. The sand that flows downhill may then increase the local slope there to beyond the critical value and so on down the sand pile. The result is called "self-organized criticality" because the system self-organizes itself to being just under the critical slope everywhere needed for the sand to move downhill. The lesson from dynamical systems here is that some systems are poised to be near their least stable configurations and thus far from their stable equilibrium.

Self-organizing criticality has been useful in understanding many different systems, including the timing and severity of earthquakes, forest fires, traffic jams, city growth, market fluctuations, and the sizes of companies (Bak, 1997; Turcotte & Rundle, 2002). What is perhaps most relevant for social psychology is that self-organizing criticality produces a characteristic signature in the statistical distribution of the fluctuations in the variables that measure the properties of a system (issue #6). That characteristic statistical distribution is called a *power law* or *fractal distribution* (Brown & Liebovitch, 2010). If this statistical distribution is observed in a social system, self-organized criticality is a possible mechanism that could have produced that self-organization.

In summary, dynamical systems have been useful in analyzing physical and biological systems to understand threshold effects, how emergent patterns self-organize, sensitivity to initial conditions, whether chance is really needed to explain seemingly random data, and the statistical properties of the variables of those systems. This suggests that the use of dynamical systems may be equally valuable in understanding similarly important issue in the social psychology of conflicts.

Summing Up and Looking Ahead

Although a relatively new perspective, the dynamical systems approach to conflict has deep roots in other theoretical orientations and research. This chapter briefly sketches the theoretical origins of the dynamical systems approach to conflict, tracing its trajectory through (1) peace and conflict studies, (2) social psychology, and (3) complexity science. Each of these traditions is valuable yet limited. The peace and conflict perspective has revealed paradoxical features of conflict and established the diverse forms that conflict can take at different levels of social reality. Within this tradition, five paradigms have been advanced to provide insight into the principles by which conflict has been studied: realism, human relations, pathology, post-modernism, and systems.

Next, we have briefly outlined six influential social psychological perspectives on conflict: game theory, social interdependence theory, social motivation theory, dual-concern theory, power dependence theory and social identity theory. Although these traditions and theories constitute rich soil for comprehending conflict, they still fail to address the key theoretical, methodological and practical challenges posed by intractable conflicts.

In the last sections of this chapter we presented recent developments in complexity science and dynamical systems, and broached the potential value of these developments for a comprehensive study of conflict. We demonstrate that dynamical systems have been useful in analyzing physical and biological systems to understand threshold effects, self-organization, emergent properties of complex systems, or sensitivity to initial conditions. We propose that the use of dynamical systems may be equally valuable in understanding important issues of social conflict. What is more, the language and concepts of this approach offer a promise of integration of existing knowledge in the fragmented field of conflict studies.

In sum, although classic and contemporary research on conflict has made great strides toward understanding the conditions and processes that determine whether conflict will take a constructive or destructive course, none of these ideas fully explain the paradoxical and diverse nature of intractable conflict. The ideas are especially wanted when it comes to the small proportion of conflicts that progress to intractability. This is where the core assumptions and insights of dynamical systems add value. This perspective has transformed the way scientists conceptualize and investigate phenomena in the physical sciences and there is reason to believe that they will do the same for the social sciences in general, and for the study of social conflict in particular. The dynamical systems approach to conflict constitutes an integrative platform for several traditions of research and theories; such a platform can promote the emergence of a unique and coherent understanding of some key features of intractable conflicts.

References

Allport, G. W. (1954). *The nature of prejudice*. Reading, MA: Addison-Wesley.

Allport, G. W. (1969). *Becoming: Basic considerations for a psychology of personality*. New Haven, CT: Yale University Press.

Alvesson, M., & Willmott, H. (1992a). Critical theory and management studies: An introduction. In M. Alvesson & H. Willmott (Eds.), *Critical management studies* (pp. 1–20). London: Sage.

Alvesson, M., & Willmott, H. (1992b). On the idea of emancipation in management and organization studies. *Academy of Management Review, 17*, 432–464.

Anderson, R. M., & May, R. M. (1991). *Infectious diseases of humans: Dynamics and control.* New York: Oxford University Press.

Axelrod, R. (1984). *The evolution of cooperation.* New York: Basic Books.

Bak, P. (1997). *How nature works: The science of self-organized criticality.* New York: Oxford University Press.

Bartoli, A., Bui-Wrzosinska, L., & Nowak, A. (2010). Peace is in movement: A dynamical systems perspective on the emergence of peace in Mozambique. *Peace and Conflict: Journal of Peace Psychology, 16*, 211–230.

Bechtoldt, M. N., De Dreu, C. K. W., Nijstad, B. A., & Choi, H. S. (2010). Motivated information processing, social tuning, and group creativity. *Journal of Personality and Social Psychology, 99*, 622–637.

Blake, R. R., Mouton, J. S., Barnes, L. B., & Greiner, L. E. (1964). Breakthrough in organization development. *Harvard Business Review, 42*, 133–155.

Boardman, S., & Horowitz, S. (1994). Constructive conflict management and social problems: An introduction. *Journal of Social Issues, 50*, 1–12.

Brodbeck, F. C., Kugler, K. G., Fischer, J. A., Heinze, J. Fischer, D. (2011). Team Level Integrative Complexity: Enhancing Differentiation and Integration in Group Decision Making. *WOP Working Paper No. 2011/2.* Retrieved from: www.psy.lmu.de/wirtschaftspsychologie/forschung/working_papers/index.html

Brown, R. (2000). Social identity theory: Past achievements, current problems and future challenges. *European Journal of Social Psychology, 30*(6), 745–778.

Brown, C. T. & Liebovitch, L. S. (2010). *Fractal analysis: Quantitative applications in the social sciences* (Vol. 165). Thousand Oaks, CA: Sage.

Burnham, T., McCabe, K., & Smith, V. L. (2000). Friend-or-foe intentionality priming in an extensive form trust game. *Journal of Economic Behavior and Organization, 43*, 57–73.

Cairns, E., & Darby, J. (1998). The conflict in Northern Ireland: Causes, consequences, and controls. *American Psychologist, 53*, 754–760.

Campbell, D. T. (1965). *Ethnocentric and other altruistic motives.* Lincoln, NE: University of Nebraska Press.

Cartwright, D. (1979). Contemporary social psychology in historical perspective. *Social Psychology Quarterly, 42*, 82–93.

Coleman, P. T. (2003). Characteristics of protracted, intractable conflict: Towards the development of a meta-framework - I. First paper in a three-paper series. *Peace and Conflict: Journal of Peace Psychology, 9*, 1–37.

Coleman, P. T. (2004). Paradigmatic framing of protracted, intractable conflict: Towards the development of a meta-framework - II. *Peace and Conflict: Journal of Peace Psychology, 10*, 197–235.

Coleman, P. T. (2011). *The five percent: Finding solutions to seemingly impossible conflicts.* New York: PublicAffairs.

Coleman, P. T., & Deutsch, M. (2001). Introducing cooperation and conflict resolution into schools: A systems approach. In D. J. Christie, R. V. Wagner, & D. D. N. Winter (Eds.), *Peace, conflict and violence: Peace psychology for the 21st century.* Upper Saddle River, NJ: Prentice-Hall.

Coleman, P. T., Hacking, A., Stover, M., Fisher-Yoshida, B., & Nowak, A. (2008). Reconstructing ripeness I: A study of constructive engagement in protracted social conflicts. *Conflict Resolution Quarterly, 26*, 3–42.

Coleman, P. T., Kugler, K. G., Bui-Wrzosinska, L., Nowak, A., & Vallacher, R. (2012). Getting down to basics: A situated model of conflict in social relations. *Negotiation Journal, 28*(1), 7–43.

Coleman, P. T., & Lim, J. (2001). A systematic approach to evaluating the effects of collaborative negotiation training on individuals and groups. *Negotiation Journal, 17*(4), 363–392.

Coleman, P. T., Vallacher, R., Nowak, A., & Bui-Wrzosinska, L. (2006). Intractable conflict as an attractor: Presenting a dynamical model of conflict, escalation, and intractability. *American Behavioral Scientist, 50*, 1454–1475.

Crocker, J., & Luhtanen, R. (1990). Collective self-esteem and ingroup bias. *Journal of Personality and Social Psychology, 58*, 60.

Darwin, C. (1859). *On the origin of species by means of natural selection, or the preservation of favoured races in the struggle for life*. London: John Murray.

Darwin, C. (1871). *On the origin of species* (5th ed.). London: John Murray.

De Dreu, C. K. W., Beersma, B., Steinel, W., & Van Kleef, G. A. (2007). The psychology of negotiation: Principles and basic processes. In A. W. Kruglanski & E. T. Higgins (Eds.), *Social psychology: Handbook of basic principles* (2nd ed.). New York: Guilford.

De Dreu, C. K. W., Beersma, B., Stroebe, K., & Euwema, M. C. (2006). Motivated information processing, strategic choice, and the quality of negotiated agreement. *Journal of Personality and Social Psychology, 90*, 927–943.

De Dreu, C. K. W., & Gelfand, M. J. (2008). Conflict in the workplace: Sources, dynamics and functions across multiple levels of analysis. In C. K. W. De Dreu & M. J. Gelfand (Eds.), *The psychology of conflict and conflict management in organizations*. New York: Lawrence Erlbaum.

De Dreu, C. K., & Van Lange, P. A. (1995). The impact of social value orientations on negotiator cognition and behavior. *Personality and Social Psychology Bulletin, 21*, 1178–1188.

De Dreu, C. K. W., & Weingart, L. R. (2003). Task versus relationship conflict, team performance, and team member satisfaction: A meta-analysis. *Journal of Applied Psychology, 88*, 741–749.

De Dreu, C. K. W., Weingart, L. R., & Kwon, S. (2000). Influence of social motives on integrative negotiation: A meta-analytical review and test of two theories. *Journal of Personality and Social Psychology, 78*, 889–905.

De Wit, F. R. C., Greer, L. L., & Jehn, K. A. (2012). The paradox of intragroup conflict: A meta-analysis. *Journal of Applied Psychology, 97*, 360–390.

De Zavala, A. G., Cichocka, A., Eidelson, R., & Jayawickreme, N. (2009). Collective narcissism and its social consequences. *Journal of Personality and Social Psychology, 97*, 1074.

Deutsch, M. (1949a). An experimental study of the effects of co-operation and competition upon group processes. *Human Relations*. Retrieved from http://psycnet.apa.org/psycinfo/1950-04051-001

Deutsch, M. (1949b). The directions of behavior: A field-theoretic approach to the understanding of inconsistencies. *Journal of Social Issues, 25*, 43–51.

Deutsch, M. (1969). Conflicts: Productive and destructive. *Journal of Social Issues, 25*, 7–42.

Deutsch, M. (1973). *The resolution of conflict: Constructive and destructive processes*. New Haven, CT: Yale University Press.

Deutsch, M. (1985). *Distributive justice: A social-psychological perspective*. New Haven, CT: Yale University Press.

Deutsch, M. (2000). A framework for thinking about research on conflict resolution training. In M. Deutsch & P. T. Coleman (Eds.), *The handbook of conflict resolution: Theory and practice*. San Francisco: Jossey-Bass.

Deutsch, M. (2002). Social psychology's contributions to the study of conflict resolution. *Negotiation Journal, 18*, 307–320.

Deutsch, M. (2006). A framework for thinking about oppression and its change. *Social Justice Research, 19*, 7–41.

Deutsch, M., & Coleman, P. T. (2000). *The handbook of conflict resolution: Theory and practice*. San Francisco, CA: Jossey-Bass.

Deutsch, M., Coleman, P. T., & Marcus, E. C. (2006). *Handbook of conflict resolution: Theory and practice* (2nd ed.). San Francisco: Jossey-Bass.

Diehl, P. F., & Goertz, G. (2001). *War and peace in international rivalry*. Ann Arbor, MI: University of Michigan Press.

Diesing, P. (1962). *Reason in society: Five types of decisions and their social conditions*. Urbana, IL: University of Illinois Press.

Disney, A., & Gbowee, L. (2012). Gender and sustainable peace. In P. T. Coleman & M. Deutsch (Eds.), *Psychological components of sustainable peace* (pp. 197–203). Berlin: Springer.

Dovidio, J. F., & Gaertner, S. L. (1986). *Prejudice, discrimination, and racism: Historical trends and contemporary approaches*. San Diego: Academic.
Druckman, D. (1994). Determinants of compromising behavior in negotiation: A meta-analysis. *The Journal of Conflict Resolution, 38*, 507–556.
Druckman, D. (2007). Negotiating in the international context. In I. W. Zartman (Ed.), *Peacemaking in international conflict methods and techniques* (pp. 111–162). New York: United States Institute of Peace Press.
Emerson, R. M. (1962). Power-dependence relations. *American Sociological Review, 27*, 31–41.
Filley, A. C. (1975). *Interpersonal conflict resolution*. Glenview, IL: Scott, Foresman.
Fisher, R. (1994). Generic principals for resolving intergroup conflict. *Journal of Social Issues, 50*, 47–68.
Fisher, R. J., & Keashly, L. (1990). A contingency approach to third party intervention. In R. J. Fisher (Ed.), *The social psychology of intergroup and international conflict resolution* (pp. 234–238). New York: Springer.
Fisher, R. J. (1997). Training as interactive conflict resolution: Characteristics and challenges. *International Negotiation, 2*(3), 331–351.
Fiske, S. (1998). Stereotyping, prejudice, and discrimination. In D. Gilbert, S. Fiske, & G. Lindzey (Eds.), *The handbook of social psychology* (Vol. 2, pp. 357–411). New York: McGraw-Hill.
Fiske, S. (2000). Stereotyping, prejudice, and discrimination at the seam between the centuries: Evolution, culture, mind, and brain. *European Journal of Social Psychology, 29*, 188–211.
Fiske, S. (2002). What we know now about bias and intergroup conflict, the problem of the century. *Current Directions in Psychological Science, 11*, 123–128.
Freud, S. (1936). *The problem of anxiety*. New York: W. W. Norton.
Freud, S. (1960). The psychopathology of everyday life. In J. Strachey (Ed.) & A. Tyson (Trans.), *The standard edition of the complete psychological works of Sigmund Freud* (Vol. 6). London: Hogarth Press (Original work published 1901).
Gerhart, B., & Rynes, S. L. (1991). Determinants and consequences of salary negotiations by male and female MBA graduates. *Journal of Applied Psychology, 76*, 256–262.
Gurr, T. R., & Scarritt, J. R. (1989). Minorities rights at risk: A global survey. *Human Rights Quarterly, 11*, 375.
Higgins, E. T. (1997). Beyond pleasure and pain. *American Psychologist, 52*, 1280–1300.
Hinkle, S., & Brown, R. (1990). Intergroup comparisons and social identity: Some links and lacunae. *Social Identity Theory: Constructive and Critical Advances, 48*, 70.
Hewstone, M., & Cairns, E. (2001). *Social psychology and intergroup conflict*. New York: Springer.
Hogg, M. A., & Williams, K. D. (2000). From I to we: Social identity and the collective self. *Group Dynamics: Theory, Research, and Practice, 4*, 81–97.
Johnson, D. W., & Johnson, R. T. (1989). *Cooperation and competition: Theory and research*. Edina, MN: Interaction Book.
Johnson, D. W., & Johnson, R. T. (2005). New developments in social interdependence theory. *Psychology Monograph, 13*, 285–360.
Jones, E. E. (1985). Major developments in social psychology during the past five decades. *Handbook of Social Psychology, 1*, 47–107.
Jones, E. E. (1998). Major developments in five decades of social psychology. In D. T. Gilbert, S. T. Fiske, & G. Lindzey (Eds.), *The handbook of social psychology* (4th ed.). New York: McGraw-Hill.
Jost, J. T., & Banaji, M. R. (1994). The role of stereotyping in system-justification and the production of false consciousness. *British Journal of Social Psychology, 33*, 1–27.
Jost, J. T., Banaji, M. R., & Nosek, B. A. (2004). A decade of system justification theory: Accumulated evidence of conscious and unconscious bolstering of the status quo. *Political Psychology, 25*, 881–919.
Karpf, F. B. (1932). *American social psychology: Its origins, development, and European background*. http://psycnet.apa.org/psycinfo/2006-04433-000/

Katz, D., & Kahn, R. L. (1978). *The social psychology of organizations* (2nd ed.). New York, NY: John Wiley.

Kegan, R. (1994). *In over our heads: The mental demands of modern life*. Cambridge, MA: Harvard University Press.

Kelley, H. H. (1967). Attribution theory in social psychology. In D. Levine (Ed.), *Nebraska symposium on motivation*. Lincoln, NE: University of Nebraska Press.

Kelley, H. H., & Thibaut, J. W. (1978). *Interpersonal relations: A theory of interdependence*. New York, NY: John Wiley.

Kelman, H. (1999). The interdependence of Israeli and Palestinian national identities: The role of the other in existential conflicts. *Journal of Social Issues, 55*, 581–600.

Kim, P. H. (1997). Strategic timing in group negotiations: The implications of forced entry and forced exit for negotiators with unequal power. *Organizational and Human Behavior Processes, 71*, 263–286.

Kim, P. H., & Fragale, A. R. (2005). Choosing the path to bargaining power: An empirical comparison of BATNAs and contributions in negotiations. *Journal of Applied Psychology, 90*, 373–381.

Kim, P. H., Pinkley, R. L., & Fragale, A. R. (2005). Power dynamics in negotiation. *Academy of Management Review, 30*(4), 799–822.

Kozlowski, S. W. J., & Klein, K. J. (2000). *A multilevel approach to theory and research in organizations: Contextual, temporal, and emergent processes*. San Francisco, CA: Jossey-Bass.

Kreisberg, L. (2005). Nature, dynamics, and phases of intractability. In C. A. Crocker, F. O. Hampson, & P. Aall (Eds.), *Grasping the nettle: Analyzing cases of intractable conflict*. Washington, DC: United States Institute of Peace Press.

Kriesberg, L. (2003). *Constructive conflicts: From escalation to resolution* (2dth ed.). Lanham, MD: Rowman and Littlefield.

Lalonde, R. N. (2002). Testing the social identity-intergroup differentiation hypothesis: 'We're not American eh!'. *British Journal of Social Psychology, 41*(4), 611–630.

Lederach, J. P. (1997). *Building peace: Sustainable reconciliation in divided societies*. Washington, DC: United States Institute of Peace Press.

Levy, J. S. (1996). Contending theories of international conflict: A levels-of-analysis approach. In C. A. Crocker, F. O. Hampson, & P. Aall (Eds.), *Managing global chaos: Sources of and responses to international conflict*. Washington, DC: United States Institute of Peace Press.

Lewin, K. (1936). *Principles of topological psychology*. New York, NY: McGraw Hill.

Lewin, K. (1947). Frontiers in group dynamics. *Human Relations, 1*, 5–41.

Lewin, K. (1948). *Resolving social conflicts*. New York, NY: Harper.

Lewicki, R. J., Saunders, D. M., Barry, B., & Minton, J. W. (2004). *Essentials of negotiation* (3rd ed.). New York: McGraw-Hill.

Liebovitch, L. S., & Scheurle, D. (2000). Two lessons from fractals and chaos. *Complexity, 5*(4), 34–43.

Livingstone, A., & Haslam, S. A. (2008). The importance of social identity content in a setting of chronic social conflict: Understanding intergroup relations in Northern Ireland. *British Journal of Social Psychology, 47*, 1–21.

Lorenz, E. N. (1963). Deterministic nonperiodic flow. *Journal of the Atmospheric Sciences, 20*, 130–141.

Luhtanen, R., & Crocker, J. (1992). A collective self-esteem scale: Self-evaluation of one's social identity. *Personality and Social Psychology Bulletin, 18*, 302–318.

Mannix, E. A. (1993). The influence of power, distribution norms, and task meeting structure on resource allocation in small group negotiation. *International Journal of Conflict Management, 4*, 5–23.

Mannix, E. A., Thompson, L. L., & Bazerman, M. H. (1989). Negotiation in small groups. *Journal of Applied Psychology, 74*, 508–517.

Mason, T. D., Crenshaw, M., McClintock, C., & Walter, B. (2007). *How political violence ends: Paths to conflict de-escalation and termination. APSA Task Force on Political Violence and Terrorism, Group 3*. Presented at the annual meeting of the American Political Science Association.

McClintock, C. (1976). Social motives in settings of outcome interdependence. In D. Druckman (Ed.), *Negotiations: Social psychological perspective*. Beverly Hills, CA: Sage.

Mooradian, M., & Druckman, D. (1999). Hurting stalemate or mediation? The conflict over Nagorno-Karabakh, 1990-95. *Journal of Peace Research, 36*, 709–727.

Morawski, J. (2000). Social psychology a century ago. *American Psychologist, 55*, 427.

Ng, S. H. (1980). *The social psychology of power*. London: Academic.

Pearce, W. B., & Littlejohn, S. W. (1997). *Moral conflict: when social worlds collide*. Thousand Oaks, CA: Sage.

Perdue, C. W., Dovidio, J. F., Gurtman, M. B., & Tyler, R. B. (1990). Us and them: Social categorization and the process of intergroup bias. *Journal of Personality and Social Psychology, 59*, 475.

Pinker, S. (2011). *The better angels of our nature: Why violence has declined*. New York: Penguin.

Pinkley Robin, L., Neale Margaret, A., & Bennett Rebecca, J. (1994). The impact of alternatives to settlement in dyadic negotiation. *Organizational Behavior and Human Decision Processes, 57*, 97–116.

Pruitt, D. G. (1981). *Negotiation behavior*. New York, NY: Academic.

Pruitt, D. G. (1997). Ripeness theory and the Oslo talks. *International Negotiation, 2*(2), 237–250.

Pruitt, D. G. (1998). Social conflict. In D. T. Gilbert, S. T. Fiske, & G. Lindzey (Eds.), *The handbook of social psychology* (4th ed.). New York: McGraw-Hill.

Pruitt, D. G., & Kim, S. H. (2004). *Social conflict: Escalation, stalemate, and settlement*. New York: McGraw-Hill.

Pruitt, D. G., & Olczak, P. V. (1995). Resolving seemingly intractable conflict. In B. B. Bunker & J. Z. Rubin (Eds.), *Conflict, cooperation, and justice: Essays inspired by the work of Morton Deutsch*. San Francisco: Jossey-Bass.

Pruitt, D. G., & Rubin, J. Z. (1986). *Social conflict*. New York, NY: McGraw-Hill.

Rahim, M. A. (1983). A measure of styles of handling interpersonal conflict. *The Academy of Management Journal, 26*, 368–376.

Rahim, M. A. (1986). Referent role and styles of handling interpersonal conflict. *Journal of Social Psychology, 126*, 79–86.

Rubin, J. Z. (1991). *Psychological approach. International negotiation. Analysis, approaches, issues. Processes of international negotiations, PIN project* (pp. 216–228). San Francisco: Jossey-Bass.

Rubin, J. Z., & Levinger, G. (1995). Levels of analysis: In search of generalizable knowledge. In B. B. Bunker & J. Z. Rubin (Eds.), *Cooperation, conflict, and justice: Essays inspired by the work of Morton Deutsch* (pp. 13–38). New York: Sage.

Rubin, J. Z., Pruitt, D. G., & Kim, S. H. (1994). *Social conflict: Escalation, stalemate, and settlement* (2nd ed.). New York: McGraw-Hill.

Saltzman, B. (1962). Finite amplitude free convection as an initial value problem – I. *Journal of the Atmospheric Sciences, 19*, 329–341.

Schelling, T. C. (1960). *The strategy of conflict*. Cambridge, MA: Harvard University Press.

Schulz-Hardt, S., Mojzisch, A., & Vogelgesang, F. (2008). Dissent as a facilitator: Individual- and group-level effects on creativity and performance. In C. K. W. De Dreu & M. J. Gelfand (Eds.), *The psychology of conflict and conflict management in organizations* (pp. 149–177). New York, NY: Taylor and Francis Group/Lawrence Erlbaum Associates.

Schwartz, I. B. (1992). Small amplitude, long period outbreaks in seasonally driven epidemics. *Journal of Mathematical Biology, 30*, 473–491.

Sherif, M., Harvey, O. J., White, B. J., Hood, W., & Sherif, C. W. (1961). *Intergroup conflict and cooperation: The robbers cave experiment*. Norman, OK: The University Book Exchange.

Sidanius, J., Levin, S., Federico, C., & Pratto, F. (2001). Legitimizing ideologies: The social dominance approach. In J. Jost & B. Major (Eds.), *The psychology of legitimacy: Emerging perspectives on ideology, justice, and intergroup relations* (pp. 307–331). Cambridge: Cambridge University Press.

Sidanius, J., Pratto, F., van Laar, C., & Levin, S. (2004). Social dominance theory: Its agenda and method. *Political Psychology, 25*, 845–880.

Sternberg, R. J., & Dobson, D. M. (1987). Resolving interpersonal conflicts: An analysis of stylistic consistency. *Journal of Personality and Social Psychology, 52*, 794–812.

Tajfel, H. (1970). Experiments in intergroup discrimination. *Scientific American, 223*, 96–102.

Tajfel, H., & Turner, J. C. (1979). An integrative theory of intergroup conflict. In W. G. Austin & S. Worchel (Eds.), *The social psychology of intergroup relations* (pp. 33–47). Monterey, CA: Brooks/Cole.

Thibaut, J. W., & Kelley, H. H. (1959). *The social psychology of groups*. New York: Wiley.

Thomas, K. W. (1976). *Thomas-Kilmann conflict model*. Tuxedo, NY: Xicom/Consulting Psychologists Press.

Tjosvold, D. (1991). *The conflict positive organization*. Reading, MA: Addison-Wesley.

Tjosvold, D., & Johnson, D. W. (1983). *Productive conflict management: Perspectives for organizations*. New York: Irvington.

Turcotte, D. L., & Rundle, J. B. (2002). Self-organized complexity in the physical, biological, and social sciences. *Proceedings of the National Academy of Sciences of the United States of America, 99*(Suppl 1), 2463–2465.

Turner, J. C., Oakes, P. J., Haslam, S. A., & McGarty, C. (1994). Self and collective: Cognition and social context. *Personality and social psychology bulletin, 20*, 454–454.

Turner, J. C., & Reynolds, K. J. (2001). The social identity perspective in intergroup relations: Theories, themes and controversies. In R. J. Brown & S. Gaertner (Eds.). *Handbook of social psychology*. Intergroup processes (Vol. 4, pp. 133–152). Oxford: Oxford University Press.

United Nations. (2004). *A more secure world: Our shared responsibility. Report of the United Nations high-level panel on threats. Challenges and change*. New York: United Nations.

Van de Vliert, E., & Kabanoff, B. (1990). Toward theory-based measures of conflict management. *The Academy of Management Journal, 33*, 199–209.

Van Lange, P. A. M., De Cremer, D., Van Dijk, E., & Van Vugt, M. (2007). Self-interest and beyond: Basic principles of social interaction. In A. W. Kruglanski & E. T. Higgins (Eds.), *Social psychology: Handbook of basic principles* (2nd ed.). New York: Guilford.

Von Neumann, J., & Morgenstern, O. (1944). *Theory of games and economic behavior*. Princeton, NJ: Princeton University Press.

Voronov, M., & Coleman, P. T. (2003). Beyond the ivory towers: Organizational power practices and a "practical" critical postmodernism. *Journal of Applied Behavioral Science, 39*, 169–185.

Waltz, K. N. (1979). *Theory of international politics* (Vol. 5). New York: McGraw-Hill.

Weingart, L. R., Bennet, R. J., & Brett, J. M. (1993). The impact of consideration of issues and motivational orientation on group negotiation process and outcome. *Journal of Applied Psychology, 78*, 504–517.

Zartman, I. W. (1989). *Ripe for resolution*. New York: Oxford University Press.

Zartman, I. W. (2000). Ripeness: The hurting stalemate and beyond. In P. Stern & D. Druckman (Eds.), *International conflict resolution after the cold war*. Washington, DC: National Academy Press.

Chapter 3
Foundations: The Dynamical Perspective on Social Processes

The preceding chapters have established the rationale for reframing the essential features of conflict in terms of the principles, metaphors, and methods of dynamical systems. As emphasized in Chap. 2, however, this reframing would be impossible without the accumulated insights and evidence provided by the study of peace and conflict and by the principles of interpersonal and intergroup experience established in social psychology. It is through the interaction of these elements—peace and conflict studies, social psychology, and dynamical systems—that the emergence of a dynamical perspective on conflict can even be envisioned, let alone developed and verified. Our aim in this chapter is to provide a broad outline of this emergent product.

We begin by describing work over the last two decades establishing the relevance of dynamical systems for the subject matter of social psychology. This approach—*dynamical social psychology*—shows how seemingly different aspects of human experience can be understood in terms of basic principles of dynamical systems in other areas of science. Beyond establishing the foundations of this perspective in dynamical systems, we acknowledge the insights generated during the early days of social psychology—well before the advent of methods and tools appropriate to identify the dynamical properties of social phenomena. In the next section, we present a basic research strategy—referred to as *dynamical minimalism*—that is designed to test hypotheses generated within the dynamical social psychology framework.

Building on this conceptual and methodological foundation, the final section illustrates how the dynamical perspective in social psychology has been employed to investigate the connection between structure and dynamics at different levels of reality—from minds to societies. Because conflict can exist at each of these levels, the approach of dynamical social psychology holds potential for exposing the invariant dynamics at work in conflict scenarios that are distinct with respect to their manifest features and that seem to reflect different processes. This chapter thus sets the stage for the discussion in subsequent chapters of important features of destructive and seemingly intractable conflict that can be reframed and investigated from the dynamical perspective.

R.R. Vallacher et al., *Attracted to Conflict: Dynamic Foundations of Destructive Social Relations*, Peace Psychology Book Series, DOI 10.1007/978-3-642-35280-5_3,

Dynamical Social Psychology

Life is complex and dynamic, even in the most banal of circumstances. Each momentary thought, feeling, or action represents a punctuation point in a continuous flow of events that interact over time, producing a complex trajectory of points that defines one's personal and interpersonal lives. To understand and appreciate human thought and behavior requires the identification of meaningful patterns in the flow of experience, while acknowledging the potential for transformation of these patterns due to the particular confluence of forces at any point in time. Surprisingly, however, an explicit recognition that temporal patterns and their potential for change represent the essence of human experience has yet to find a secure home in mainstream social psychology. For want of appropriate tools—and for reasons of convenience—the approach of early dynamical theorists such as James, Mead, and Lewin has been replaced by an approach to personal and interpersonal processes essentially stops the clock to isolate static features of mind and action that are tethered to a small number of external causes.

The field of psychology is undergoing a transformation, however, with the emergence of a new paradigm for psychology that pays homage to the complexity and dynamism of human thought and behavior (cf. Guastello, Koopmans, & Pincus, 2009; Vallacher & Nowak, 1997, 2007). A wide range of topics—including cognitive neuroscience (cf. Port & Van Gelder, 1995), developmental psychology (e.g., van Geert, 1991; Smith & Thelen, 1993; Thelen & Smith, 1994), organizational behavior (Axelrod & Cohen, 2000; Axelrod, Riolo, & Cohen, 2002; Guastello, 1995), and political sociology (e.g., Axelrod, 1984)—have been reframed in recent years in ways that allow insight into basic dynamic properties. Dynamical social psychology epitomizes this new approach to theory construction and research. Its goal is to reframe the topical landscape of social psychology—from micro processes within the individual to macro processes within social systems—in terms of basic principles of dynamical systems and complexity science (cf. Nowak & Vallacher, 1998; Vallacher & Nowak, 1994a, 1997, 2006, 2007).

The Added Value of the Dynamical Perspective

The dynamical approach differs from the traditional approach in two fundamental ways. In the dynamical approach, first of all, an external force does not simply "cause" a one-step response in a person or group, but rather interacts with the internal dynamics of the person or group to trigger a process that unfolds over time. Consider, for example, the different scenarios likely to unfold when kicking a chair (an inanimate object) versus kicking a dog (a very animate object). Although the chair will move proportionately in response to the kick, the dog's response will reflect features of its internal state (e.g., fear, anger, confusion). Thus, the same cause (the kick) might provoke running away (fear), biting (anger), or a frozen posture (confusion). Beyond that, the immediate response to an external force may diminish, perhaps rapidly or perhaps slowly, or it may become intensified—again, perhaps rapidly or slowly. A member of a group who is insulted or humiliated at time 1, for example, may experience anger or intimidation at time 2 (e.g., right away) but over time this experience may evolve in completely different directions—intensified anger with an increasing desire for retaliation or, conversely, diminished anger replaced with a sense of inferiority that warrants humiliation. For that matter, the person's experience may oscillate over time between these divergent emotional states. Thus, in focusing only on a response at some arbitrary (but convenient) point in time, the traditional approach may produce results and conclusions that are uninformative, perhaps even misleading.

The second distinguishing feature of dynamical social psychology is the recognition that external forces rarely operate in isolation. Instead, they interact with each other in complex ways to promote a particular trajectory of thought, emotion, and behavior. Reducing the playing field to a small number of factors may make sense from a pragmatic point of view—after all, no one wants to decompose the statistical interaction among more than three independent variables. But in principle, the effect of a set of variables may change dramatically depending on the presence of other relevant variables in a particular context. The particular trajectory of thoughts and feelings generated by an insult, for example, may differ dramatically depending on the salience of an untold number of other factors—the way in which the insult is delivered, the relationship between the "insulter" and "insultee," the self-esteem or current mood of the insultee, the presence or absence of other people, and so forth. The interaction among these factors, moreover, is not simply a combination of their separate influences but rather may represent an "emergent" phenomenon that would not occur if any of the factors were not operative.

Together, the emphasis on temporal patterns and the interplay of factors that shape and influence these patterns represents a transformation in the way social psychological phenomena are conceptualized and investigated. We hasten to add, however, that the dynamical approach to social psychology actually has deep roots. For one, it borrows heavily on ideas and methods developed in the natural sciences under the rubric of dynamical systems and complexity (cf. Haken, 1978; Holland, 1995; Johnson, 2001; Schuster, 1984; Strogatz, 2003; Weisbuch, 1992). But dynamical social psychology can also be traced, ironically, to the very earliest days of social psychology—well before the advent of nonlinear dynamical systems in the natural sciences.

Roots in Classic Social Psychology

In important respects, the emergence of dynamical social psychology is really the re-emergence of ideas, metaphors, and integrative themes developed during the formative years of social psychology. This period provided important and enduring insights into the dynamic nature of human experience in various domains (e.g., cognition, interpersonal behavior, group dynamics), based on the recognition that psychological states are subject to processes of change and transformation. These insights provided the foundation for theories and research paradigms developed decades later—including the theories and paradigms emphasizing dynamism and complexity that have emerged in recent years.

The Founding Fathers of Social Psychology

Social psychology is a young discipline, dating back scarcely more than a century. The insights and themes that provided the inspiration for the field are associated with some of the most prominent figures in the history of psychology. And these figures were keenly sensitive to the dynamism and complexity of human experience. Near the end of the nineteenth century, for example, William James (1890) coined the term "stream of consciousness" to capture the endless cascade of thoughts, memories, images, and fantasies that compete for attention on a moment-to-moment basis. During this same period, Cooley (1902) developed a model of motivation that emphasized people's constant press for action, even in the absence of external forces and incentives. For Cooley, people were never at rest in a psychological sense, but rather experienced motivation from varying sources that shifted in their relative salience on a continuous basis.

Comprehensive theories were also advanced during this time that reflected the dynamic underpinnings of mind and action. Psychodynamic theories (e.g., Freud, 1920/1964, 1937), for example, held that overt behavior and conscious thought represent the "tip of the iceberg," masking an unconscious and persistent struggle to resolve conflicting motivational forces internalized by the person during his or her formative years. George Herbert Mead (1934), meanwhile, emphasized the dynamic nature of conscious thought in his model of *symbolic interactionism*. The focus of his theory was people's capacity for symbolic representation and the enormous range of interpretation of "objective reality" (e.g., interpersonal behavior, social and natural events, one's personal characteristics) to which this capacity gives rise.

Somewhat later, Solomon Asch (1946, 1956) emphasized the emergent nature of mental and affective states. Social judgment, for example, was said to reflect the dynamic interplay of specific thoughts and feelings, with this interplay promoting a unique Gestalt that is not reducible to the additive components of the elements themselves. And in one of the earliest social psychology textbooks, Krech and Crutchfield (1948) framed intrapersonal and interpersonal processes in terms of Gestalt psychology, with an explicit emphasis on the constant reconfiguration of thought and behavior in response to conflicting fields of psychological forces.

The Legacy of Kurt Lewin

Dynamical social psychology, however, owes its biggest debt to Kurt Lewin. Indeed, Lewin is widely regarded as the father of "dynamic psychology." Like the founding fathers who preceded him, Lewin (1938, 1948a) felt that dynamic processes provided the key to understanding human thought and behavior in social contexts. But Lewin was trained as a physicist and brought to bear his appreciation for physical dynamics when theorizing about mental, behavioral, and social dynamics. Had the tools now available for assessing dynamic properties been available during his professional lifetime, he might well have established what we only recently have come to know as dynamical social psychology.

Central to Lewin's theorizing was the idea of *dynamic field*, which refers to the complex constellation of forces that determine a person's movement toward or away from a particular state or course of action. Any factor having an effect on a person in a given context could be included in the person's dynamic field. In incorporating a wide range of potential influences on thought and behavior, the concept of dynamic field allows for the integration of different disciplines (e.g., biology, economics, politics, culture) and levels of analysis (e.g., from hormones to social norms). These factors, moreover, could act in varying degrees of concordance versus opposition and thus promote complex trajectories of mind and action. When the forces tend to point in the same direction, for example, the person's trajectory might look goal-directed with an unambiguous orientation toward achieving some state (e.g., food, resources, fame, intimacy). When the salient factors in the dynamic field point in opposite directions, the person's trajectory might conform to ambivalence or other manifestations of mental and behavioral conflict. For Lewin, simply focusing on a person's current state without looking at the system of forces in which it is embedded provides incomplete, if not misleading, insight into the field of forces experienced by the person.

This is not to say that a person cannot stabilize on a given state (e.g., a judgment or mood) or action orientation (e.g., approach vs. avoid, cooperation vs. competition). To the contrary, Lewin (1948a) emphasized the *equilibrium* tendencies of mind and action, such that people tend to converge on a relatively stable way of thinking and behaving. In dynamic terms, there are two routes to equilibrium. Equilibrium is attained, first of all, if all the forces in the dynamic field point in the same direction and thus promote a single state or action orientation. Though the absence of conflicting forces is probably a rare event in the multi-faceted contexts of everyday life, one can conceive of decisions, judgments, and behaviors that reflect a single-minded view free of contradictions. Indeed, social cognition research since Lewin's time has identified several processes—selective exposure to information, confirmatory bias, biases in interpretation and inference, and so on—that can promote and maintain an unequivocal mental and action orientation (e.g., Jones & Gerard, 1967; Fiske & Taylor, 1991; Wegner & Vallacher, 1977). And as we shall see elsewhere in this book, our own work has identified important factors in conflict scenarios that promote a collapse of complexity in the dynamic field and thereby bring about rigidity in interpersonal and inter-group relations.

Equilibrium can also be reached when the dynamic field consists of conflicting factors and influences. This occurs when there is a balance among the competing forces and tensions in the system, so that movement in one direction (e.g., approach) is countered by movement in the opposite direction (e.g., avoid). In this view, a person might feel both drawn to and repulsed by a particular course of interpersonal behavior—a desire for intimacy matched by an equally strong concern with autonomy, for example—and stabilize at the point where these forces are equal (e.g., involvement with an air of detachment). Lewin noted, however, that when the opposing forces are not only equal, but also strong and felt intensely, the person is likely to demonstrate departures in both directions from the equilibrium point on some timescale (e.g., seconds, days, or weeks). In an intensely felt approach-avoid conflict, for example, a person might show noteworthy oscillations in his or her thoughts and actions, with the equilibrium point providing an unstable average of the overall trajectory.

Equilibrium tendencies are largely adaptive because they allow for a stable frame of reference amidst the continuous flow of information and events in daily life. But equilibrium can also be manifest as rigidity in thought and behavior, with the natural tendency toward accommodation and change suppressed in favor of maintaining a stable and coherent orientation. In a complex and ever-changing world, such rigidity maintained over an extended period of time can promote an increasingly out-of-touch mindset and generate a dysfunctional relation to social reality. For Lewin, a central challenge of social life is achieving a balance between stability and change—between the maintenance of a coherent perspective and the readiness to make adjustments in response to a changing field of forces.

Lewin (1948b) proposed a three-stage process for inducing change in a system that has developed a stable equilibrium characterized by rigidity. The first stage—"unfreeze"—involves disassembling the global mindset and action orientation. Unfreezing requires rechanneling the dynamic forces acting on the system so that they can no longer position the system at only one point. The second stage—"move"—involves getting the person to reconsider the relevant field of forces acting on him or her and consider new ways of integrating them. This seems like a very reasonable thing for the person to do, but Lewin argued that it was impossible to accomplish without first unfreezing the system so that the person is motivated to reestablish a stable equilibrium. The process of "moving" is inherently unstable and thus represents a transitory stage in mental and behavioral dynamics. The third stage—"freeze"—plays on this instability and the person's desire to regain a stable orientation. In effect, the person embraces a new equilibrium that provides coherence for the field of forces, if that is possible, or achieves a new equilibrium that represents a new balance among conflicting forces.

The "unfreeze—move—freeze" scenario is easy to appreciate in principle, but it can prove difficult to implement in practice. Recent models of mind and action, however, provide means by which a stable structure can be deconstructed—or unfrozen, in Lewinian terms—and reconfigured in line with a very different equilibrium (e.g., Davis & Knowles, 1999; Vallacher & Wegner, 2012). As we shall see elsewhere in this text, in fact, this general scenario and its means of implementation provide a means by which intractable conflicts—interpersonal or inter-group relations characterized by rigid hostility—can be transformed into benign or even positive social relations.

The Trajectory of Classic Social Psychology

Despite the focus on dynamism and complexity in classic social psychology, in subsequent years these defining features of human experience were commonly downplayed in theory and research. The de-emphasis of dynamism is apparent in theory and research on the cognitive bases of social behavior. So although James (1890) underscored the dynamic nature of conscious experience in his stream of consciousness metaphor, cognitive models in social psychology for much of the twentieth century tended to portray the development of stable attitudes, causal attributions, categories, beliefs, and other mental structures as a one-step response to external influence rather than as a process that unfolds over time due to people's internal dynamics. Equilibrium tendencies provided the focus of attention, in other words, with little attention given to the dynamic processes that made equilibrium attainment possible. Stability is clearly central to human thought and behavior, but focusing primarily on this feature of human psychology overlooks half the picture.

The oversight of complexity, meanwhile, is apparent in many models of social behavior developed in the middle decades of the twentieth century. Lewin (1938, 1948a) offered a comprehensive theory that framed human experience in terms of a field of ever-present and conflicting forces, but scholars trained in this tradition tended to reduce the playing field to a small subset of such forces, with the vast majority of them removed from the equation and thereby "controlled" through randomization procedures. At best, researchers tended to look for 2-way or perhaps 3-way interactions among "independent variables," ignoring the possibility that such interactions were embedded in a much larger field of relevant forces. The essence—and challenge—of social experience is achieving a more-or-less stable, yet flexible equilibrium despite the multitude of conflicting forces to with individuals and groups are exposed on a continual basis.

Roots in Dynamical Systems

In recent years, social psychology has shown signs of coming full circle, returning to the deep intuitions concerning personal and interpersonal processes that were articulated by the field's founding fathers. This re-emergence of appreciation for complexity and dynamism was made possible by developments in the understanding and investigation of nonlinear dynamical systems in the 1970s and 1980s, and by the application of these developments to social psychological phenomena within the last two decades. These advances have enabled theorists and researchers to explore the inherent dynamism and complexity of human experience, while recognizing the tendencies toward stability and simplicity characterizing different domains of functioning. The dynamical perspective, in fact, is *defined* in terms of these opposing aspects of human experience, and has begun to construct testable theories that capture the interplay of dynamism and stability at different levels of social reality—from individual thoughts and feelings to group processes and intergroup relations.

In its most basic sense, a dynamical system is simply a set of interconnected elements that evolve in time. The evolution of the system is due to the elements influencing each other to achieve a coordinated state that characterizes the system as a whole. The elements comprising a system can represent anything from atoms and molecules to animals and planets. The task of theory construction is to specify how the elements interact with each other over time to promote properties and behavior at a higher level of functioning. In physics and chemistry, a fundamental concern is how subatomic particles interact to produce visible matter and forces; in ecology, the concern is how animals interact to generate and maintain a balance between predator and prey; in cosmological theory, the focus is how celestial bodies such as planets and moons influence each other to produce stable orbits.

Common to otherwise distinct systems is the phenomenon of *emergence*. In general terms, this simply means that the higher-order property or behavior that results from the mutual influence among elements cannot be reduced to the properties of the elements. Indeed, one of the most important discoveries of modern science is that complex properties on the system level may emerge from simple properties of the system's elements and simple rules describing the interaction among the elements (e.g., Haken, 1978; Holland, 1995; Johnson, 2001; Kauffman, 1995; Waldrop, 1992; Wolfram, 2002). Very simple rules describing the interaction of neighboring cells, for example, can reproduce the patterns of pigmentation observed in animals, the shapes of plants and shells (e.g., Meinhardt, 1995; Wolfram, 2002), and the arrangement of columns in the brain's visual cortex (Miller, Keller, & Stryler, 1989). Because higher-level properties and behaviors emerge from the internal workings of the system, the process is commonly referred to as *self-organization*.

Emergence by means of self-organization is not restricted to physical and biological systems, but can be observed in social systems as well. Durkheim (1938), for instance, proposed that new properties exist at the group level that cannot be reduced to properties at the level of individual actors. The group-level properties—norms, customs, ideologies, and the like—emerge from the social interaction of individuals with very basic concerns, such as exchanging information, passing judgments on recent events, or impressing one another. Within psychology, the emergence of complex patterns at the system level from the operation of simple rules at the level of system elements has been demonstrated in both cognition (cf. Port & Van Gelder, 1995) and social behavior (e.g., Nowak & Vallacher, 1998; Read & Miller, 1998; Vallacher, Read, & Nowak, 2002). Unacquainted individuals walking on a crowded city street, for example, are only concerned with getting to their respective destinations in the shortest period of time, and doing so without bumping into one another. With repeated iteration over time, however, these simple rules of self-interest promote the emergence of an organized movement pattern in the crowd as a whole that is remarkably similar to patterns observed in the study of fluid dynamics.

Once a higher-level state emerges by means of self-organization, it constrains the behavior of the elements that give rise to it. When a stable set of orbits emerges in a planetary system, for example, new elements (e.g., a passing asteroid or comet) are captured by the orbit structure and lose their original trajectory. Because the emergent system-level property or behavior "attracts" the behavior of both existing and new elements, it is referred to as an *attractor*. A system's attractor thus stabilizes the system and actively resists change resulting from outside influences. An influence may perturb the system, but this disruption of the system's dynamics is likely to be short-lived. An attractor, in other words, is a state or pattern of changes to which the dynamics of a system converges over time.

If change does occur, it is because the pattern of mutual influences among the elements has been weakened, undermining their coordination and thereby weakening the coherence and stability of the higher-order state. From this disassembled state of affairs, however, the system is primed for self-organization and the emergence of a new higher-order state that provides a different dynamic configuration of the lower-level elements. Over time, then, dynamical systems tend to display periods of stability and resistance to change, punctuated by periods of disassembly that set the stage for a new round of self-organization and emergence. The parallel to Lewin's "unfreeze—move—refreeze" scenario, articulated in the first half of the twentieth century, is hard to miss.

Dynamical Minimalism

Human social behavior is undeniably complex. One might reasonably assume that such complexity results from the interaction of a large number of principles or rules that specify how factors in social life shape the dynamics of mind and action. In this view, a simple explanation involving a small number of operating rules is inadequate to account for the enormous diversity in people's intrapersonal and interpersonal lives. It would seem, then, that theory construction in social psychology is faced with a dilemma: one can advance a simplistic (and parsimonious) theory but miss the true essence, not to mention the nuances, of the phenomenon, or one can develop a complex theory involving many rules but wind up with an explanation that is as complex and multi-faceted as the phenomenon itself.

This dilemma is based on a false premise. Indeed, one of the most fascinating—and seemingly paradoxical—insights forthcoming from the study of nonlinear dynamical systems is that extremely complex phenomena can be understood with recourse to a small number of very simple rules (cf. Schuster, 1984; Wolfram, 2002). The approach to social processes based on this insight is referred to as *dynamical minimalism* (Nowak, 2004). This research strategy is *minimalist* in that it attempts to identify the most rudimentary rules in governing a system's behavior. It is *dynamical* in that it assumes that the system's behavior evolves in time through repeated interactions among these simple rules. The goal of dynamical minimalism, in effect, is to achieve parsimony in theory construction, but to do so without stripping the phenomenon of its subtlety and nuance.

Levels of Description and the Role of Computer Simulation

Dynamical minimalism offers a new way of thinking about the relation between micro and macro levels of description. Traditional models of social processes commonly assume reductionism, such that the rules observed at one level of description correspond to the rules operating at another level. When this assumption is translated into practice, the properties at a macro level of description are framed in terms of analogous properties of the elements comprising lower levels of description. To explain the relation between poverty and crime in a social system, for example, a theorist might reduce this relation to the relation between frustration and crime at the level of individuals in the social system.

The dynamical approach to social processes does not assume isomorphism among levels of description. To the contrary, the rules specifying the interaction among a system's elements are likely to generate very different rules at higher levels of system behavior. This idea is central to the concept of emergence, discussed earlier. Emergence sounds great in principle and has intuitive appeal, but it represents a paradox for theory construction. If properties at a macro level cannot be derived from properties of the system's lower level elements, how can knowledge of the lower-level elements generate an explanation of the higher-level properties?

The use of computer simulations resolves this apparent paradox. In computer models, one can specify the properties of system elements and the rules of interaction among these elements. As the elements interact over time in accordance with these rules, dynamic properties appear at the system level that were not programmed, let alone assumed, for the elements themselves. With computer simulations, then, a theory constructed at a basic level of psychological reality (e.g., moment-to-moment thought process, dyadic social interaction) can be tested at a higher level of psychological reality (e.g., social judgment, group norms). In so doing, a larger theory depicting the link between micro- and macro-levels of social reality is generated.

Computer simulations are essential in the dynamical minimalism approach for another reason. The basic elements comprising a system are often uninteresting, even trivial, and many of the interactions among them may turn out to have only minor impact on the systems' global properties. Some properties of system elements, however, may appear to be uninteresting or trivial, but the interactions among them over time may have significant impact on the system's higher-order properties. Prior to the construction and testing of the computer model, it may not be obvious which properties are trivial and which are essential for the emergence of higher order properties. Dynamical minimalism is designed to make this distinction and to construct a model that incorporates only the properties that are critical for the emergence process. In particular, one can systematically vary the assumptions regarding the properties of elements and their interactions, and then observe which assumptions promote meaningful changes at the macro level. Those properties that prove to have trivial consequences at the macro level can then be eliminated from the model. With this general strategy, computer simulations allow distillation of the minimal set of features (elements and rules of interaction among elements) necessary to capture the essence of a phenomenon.

Dynamical Minimalism in Perspective

Theory construction in the social sciences faces difficult choices. Because the subject matter of human social experience is highly complex and nuanced, attempts to describe it in simple terms runs the risk of trivializing the phenomena in question. On the other hand, the time-honored principle of parsimony requires that the simplest theory capable of explaining a phenomenon should be accepted, even if other theories appear "deeper" and more nuanced. Confronted with this dilemma, the theorist seems constrained to find a trade-off between the complexity of social experience and the demands for theoretical simplicity. Dynamical minimalism resolves the apparent contradiction between parsimony and complexity as well as that between reduction and emergence. In the dynamical approach, simple rules describing the basic levels of a social process may be sufficient to promote the emergence of very complex structures at a global level of the process. A simple explanation can be advanced and verified, then, without forfeiting deep and comprehensive understanding of a phenomenon of interest.

Beyond that, dynamical minimalism holds promise for providing coherence to a highly fragmented field. Such integration is possible because the dynamical minimalism approach identifies formal principles that cut across topical boundaries. Two processes may have very different surface features, but the underlying rules leading to the emergence of their properties may be the same. Public opinion—a collective process—and self-concept—a personal process—are clearly different phenomena, for example, but research conducted within the dynamical minimalism paradigm has established that remarkably similar rules underlie the emergence of both (see Nowak, Szamrej, & Latané, 1990; Nowak, Vallacher, Tesser, & Borkowski, 2000). In providing a means of integrating different topics at different levels of psychological reality, dynamical minimalism has the potential for creating a foundational science for the social sciences. And because this approach is based on insights forthcoming from the study of nonlinear dynamical systems in the natural sciences, it may provide the basis for an explanation for human experience that is grounded in what we know about all other aspects of the world around us.

We hasten to add that the use of computer simulations, although central to dynamical minimalism, cannot alone verify a theoretical model. The use of computer simulations is critical in identifying the properties that are central to the model and investigating the consequences of these properties for the higher order functioning of the system under investigation. Once these consequences have been identified, they provide the basis for hypotheses that need to be tested in empirical research. Empirical verification may take the form of laboratory experiments with appropriate controls, but it can also take the form of archival analyses that make use of data bases, historical records, and anthropological methods of data collection. Sometimes, too, the relation between computer simulations and empirical verification is reversed. Thus, data collected in laboratory studies or through archival analyses can suggest the basic properties and rules of a model, which are then implemented in a computer model to test whether the anticipated emergent properties are observed. The results of such simulations, in turn, can provide the bases for new hypotheses to be tested empirically, and so on, in an iterated approach that holds potential for generating progressive understanding of the phenomenon in question. The reciprocal feedback loops over time among theory construction, computer simulations, and empirical data are central to the dynamical minimalism approach.

Dynamical Social Psychology in Action

Every aspect of human experience is characterized by the complementary tendencies of stability and change. The mind is in constant motion, generating an endless moment-to-moment flow of thoughts and feelings, yet it also demonstrates strong equilibrium tendencies that provide coherence and stability in social judgment. Social interactions unfold with a continuous flow of words and gestures that is never replicated exactly, yet they admit to regularities and patterns. Relationships are defined in terms of the evolution of roles and sentiments, yet they conform to scripts and agendas that constrain the actions of even the most intimate partners. The challenge of dynamical social psychology is how best to capture the interplay of sustained dynamics and stabilization tendencies in a way that both appreciates the unique character of each realm and yet identifies common principles that promote an integrative understanding of human experience. The basic principles of dynamical systems theory are tailor made for this task. In the following sections, we illustrate how these principles map onto the unique subject matter of social psychology.

Intrinsic Dynamics

People's thoughts, judgments, feelings, social interactions, and social relations evolve and change in the absence of external influence. The internally generated nature of psychological processes was central to the early formulations of social psychology, and it has self-evident intuitive appeal for laypeople. Contemporary social psychological research, however, rarely focuses on the internally generated dynamics of personal and interpersonal phenomena (cf. McGrath & Kelley, 1986; Nowak & Vallacher, 1998). Instead, research typically concentrates on the prediction of outcome measures (dependent measures) from the knowledge of other factors (independent variables). External causation is certainly relevant to social processes, and the focus on outside forces has generated important insight into social psychological phenomena. But external factors do not exert their effect by acting on an empty or passive system. Rather, they interact with the *intrinsic dynamics* associated with the process in question. The centrality of intrinsic dynamics is observable at different levels of social reality, from basic intrapersonal processes to macro-level societal phenomena.

Mental Dynamics

When people think about and evaluate issues, objects, and other people, they do not simply have one thought in mind that carries the day. Rather, as James (1890) famously noted in his "stream of thought" metaphor, mental process is characterized by an ever-changing succession of concerns, reflections, fantasies, memories, insights, and stray thoughts. To an appreciable extent, these cognitive and affective elements become progressively organized into higher-order thoughts and judgments that become less variable and erratic. Simply thinking about an attitude object (e.g., another person) in the absence of any external influence, for example, tends to promote more extreme (polarized) evaluations of the object over time as the inconsistencies among specific thoughts are eliminated (Tesser, 1978).

Research within the dynamical perspective has identified temporal patterns of thought that are considerably more elaborate than attitude polarization (Vallacher & Nowak, 1994b). These patterns, moreover, often provide a better characterization of a person's mental make-up than do the summary aspects of the person's mental process (e.g., overall attitude, final decision) that are more commonly the focus of investigation. A judgment that is neutral (i.e., neither highly positive nor highly negative) when collapsed over time, for instance, can have very different meanings and implications, depending on the intrinsic dynamics of the judgment process (Vallacher, Nowak, & Kaufman, 1994). When neutrality reflects relatively little variation in evaluation occurring on a relatively slow time scale, the summary judgment might indeed reflect a truly neutral sentiment (or detachment). But if neutrality reflects oscillation between highly positive and highly negative judgments on a rapid timescale, the summary judgment signifies heightened involvement and ambivalence rather than neutrality per se. In this case, characterizing a person's thoughts and feelings in terms of an average value collapsed over time would be highly misleading. A person's feelings toward an intimate partner, for example, may reliably alternate between love and hate; averaged over time, one might conclude that the person feels neutral toward the partner—a feeling that is never actually experienced.

Action Dynamics

Actions typically have a hierarchical structure, in that the performance of an action entails the coordinated interplay of more basic actions. The act of "going to work," for example, may involve getting dressed, leaving the house, driving a car, parking the car, and entering a building. Each of these lower-level acts can, in turn, be decomposed into yet more basic lower-level elements. "Driving," for example, consists of starting the car, turning the steering wheel, making turns, and braking. Each level in an action hierarchy has a different time scale, with the lower-level acts taking place in shorter intervals of time (Newtson, 1994). "Going to work" occurs on a longer time scale than does "driving," for example, and the time scale for "driving" is longer than that for each instance of "turning the steering wheel." The intrinsic dynamics of action thus span the levels of action in an overall action hierarchy. A person's behavior may look like a continual succession of momentary movements when defined in low-level, mechanistic terms but take on the appearance of switching between qualitatively different actions, each occurring on a longer time scale, when defined in higher-level terms.

Research in the framework of action identification theory (Vallacher & Wegner, 1987, 2012) has demonstrated that when a person is induced to think about his or her action in lower-level terms, he or she is predisposed to the emergence of higher-level action understanding. In action identification theory, emergence takes place when the person is exposed to cues (e.g., social feedback by an expert or a peer) that suggest a higher-level meaning for the action. From the perspective of dynamical systems, emergence can also occur in the absence of external cues as the person reflects on his or her action, presumably because the lower-level action elements self-organize into a higher-level act identity that provides subjective integration for these elements (Vallacher & Nowak, 1997). Consider, for example, an unfamiliar social setting in which a person is inclined to identify his or her action in relatively low-level terms—for example, as "smiling," "talking to strangers," and "moving around the room"—without a clear sense of what he or she is doing in more meaningful terms. This low-level state renders the person vulnerable to social feedback regarding his or her social presence (e.g., likeable versus unlikeable, charming versus boring) that provides an avenue of emergent understanding. In the absence of such feedback, meanwhile, the person may experience emergent understanding as the lower-level identities become integrated by means of self-organization into a coherent sense of what he or she was doing.

Interaction Dynamics

Social interaction involves the coordination of individuals' behavior over time. Considerable research has focused on the interpersonal coordination of relatively low-level actions, such as speaking (e.g., Condon & Ogston, 1967; Dittman & Llewellyn, 1969) and limb movement (e.g., Kelso, 1995; Newtson, 1994; Turvey, 1990). In one line of research, for example, two people are asked to swing their legs while sitting down across from one another, with one person swinging his or her legs in time to a metronome and the other person trying to match those movements (Beek & Hopkins, 1992).

This research has revealed two forms of social coordination: *in-phase*, with the individuals swinging their legs in unison, and *anti-phase*, with the individuals swinging their legs with the same frequency but in the opposite direction. Individuals can maintain anti-phase coordination only up to a certain frequency of movement, at which point they switch to in-phase coordination. When the frequency is then decreased, at some value they are able to coordinate anti-phase again, but this tempo is significantly lower than the point at which they originally began to coordinate in-phase. Each pattern of behavior (in-phase and anti-phase), in other words, is maintained despite incremental changes in movement frequency, until a threshold of frequency is reached, beyond which there is a rapid and qualitative change to the other behavior pattern (Kelso, 1995). This phenomenon—termed hysteresis—indicates that movement coordination can be analyzed as a nonlinear dynamical system. Modes of movement coordination more complex than in-phase and anti-phase have also been identified in this general research paradigm (see Baron, Amazeen, & Beek, 1994; Turvey, 1990).

Interpersonal dynamics are not limited to the coordination of motor movements, but also include the coordination of higher-level actions (e.g., goals, plans) and internal states (e.g., emotions, judgments). There is evidence that the quality of a social relationship is reflected in the ability of partners to coordinate in-phase with respect to their respective higher-level actions, opinions, and feelings (e.g., Baron et al., 1994; Guastello, Pincus, & Gunderson, 2006; McGrath & Kelley, 1986; Nowak, Vallacher, & Zochowski, 2005; Tickle-Degnen & Rosenthal, 1987). More generally, the ebb and flow of action, feelings, and information exchange may convey deeper insight into the nature of a relationship than might global indices such as the summary action tendencies, the average sentiment, or the amount of information exchanged. This captures the lay intuition that people who feel positively about one another are "in synch" or "on the same wavelength" with respect to their internal states.

Societal Dynamics

Research suggests that knowing the temporal trajectory associated with the emergence of norms and public opinion may provide greater insight into the society's future make-up and likely response to external forces (e.g., threats) than simply knowing what the societal norms and opinions are at a single point in time (Nowak et al. 1990). When norms and opinions develop gradually over a long period of time, for example, the society shows resistance to external threats or even to new information that might promote better economic conditions. However, societal change in political and economic ideology can also occur in a rapid, nonlinear manner (e.g., Nowak & Vallacher, 2001), with a temporal trajectory that resembles phase transitions in physical systems (Lewenstein, Nowak, & Latané, 1993). Societies undergoing such nonlinear transitions, moreover, are vulnerable to subsequent rebounds of the earlier ideologies and highly responsive to threats and new information, and they may experience a period of sustained oscillation between conflicting worldviews (Nowak & Vallacher, 2001).

This nonlinear scenario was manifest in several Eastern European countries after the collapse of communism in the late 1980s. The communist regimes suffered a humiliating defeat during this period, promoting a dramatic switch in public opinion to pro-democratic parties. This switch reversed itself in the next election cycle, however, as people experienced economic setbacks due to the destabilization of existing social and political structures. This rebound was then reversed in subsequent election cycles, and so on, in a pattern of change characterized by dramatic oscillations in political and economic worldviews. The nonlinear nature of social change mirrors basic principles of nonlinear dynamical systems (cf. Latané, 1981; Lewenstein et al., 1993) that have been implemented in computer simulations (cf. Nowak & Vallacher, 1998, 2001; Nowak et al., 1990).

Computer simulations are invaluable in this research because they can track the long-term consequences of basic dynamical processes. Changes in a society that might occur over years or even decades can be compressed into the time it takes to run a computer program. Computer simulations are also valuable because they can visualize the emergence of social structures due to dynamic processes (e.g., Latané & Nowak, 1997). The simulation of social change in Eastern Europe, for example, showed the emergence of spatial "clusters" (corresponding to geographical regions) of public opinion that was observed in these countries (Nowak & Vallacher, 2001).

Attractor Dynamics

Psychological systems display intrinsic dynamics, but they also demonstrate stability and remarkable resistance to change. Each day, people encounter vast amounts of information relevant to social judgment and interpersonal relations, with much of this information being mutually contradictory. Yet, people typically manage to form and maintain relatively coherent patterns of thought and behavior in their everyday lives. Two people in a romantic relationship, for example, are likely to experience a wide variety of thoughts and feelings about one another, but over time each person's mental state will tend to converge on positive sentiment toward the other. So despite the ever-changing nature of intrapersonal and interpersonal experience, people's mental, affective, and behavioral states tend to converge on relatively narrow sets of specific states or on patterns of change between such states. These states or patterns of change between states represent *psychological attractors*.

Psychological Attractors

When a system is at its attractor, it tends to maintain that state despite forces and influences that have the potential to destabilize it. An external influence may perturb the system and move it to another state, but the system will return fairly quickly to its attractor. Several well-documented psychological phenomena imply the existence of an attractor. Self-regulation, for example, is defined in terms of impulse control, resistance to temptations and distractions, and the maintenance of internal states representing personal standards and values (cf. Carver & Scheier, 1999; Vallacher & Nowak, 1999). In like manner, psychological reactance (Brehm & Brehm, 1981), self-esteem maintenance (Tesser, Martin, & Cornell, 1996), and self-verification (Swann, 1990) each represent the tendency of mental systems to converge on a particular state (e.g., a level of self-esteem) and to resist outside forces that threaten to dislodge the person's judgments and beliefs from that state.

It is important to note, however, that attractors are not limited to goals, intentions, or other desired states. For example, a person might display a consistent pattern of antagonistic social behavior despite concerted efforts to avoid behaving in this manner. In similar fashion, a person with low self-esteem may initially embrace flattering feedback from an acquaintance, but over time the person is likely to discount or reinterpret this feedback, displaying instead a pattern of thought that converges on a negative self-evaluative state (e.g., Swann, Hixon, Stein-Seroussi, & Gilbert, 1990). In intergroup relations, meanwhile, warring factions may exhibit conciliatory gestures when prompted to do so but revert to a pattern of antagonistic thought and behavior when outside interventions are relaxed (Coleman, Vallacher, Nowak, & Bui-Wrzosinska, 2007). When a system's dynamics are governed by an attractor, in other words, the system will consistently evolve to a particular state, even if this state is not hedonically pleasant and will return to this state despite being perturbed by forces that might promote a more pleasant state.

Multiple Attractors

A psychological system can have more than one attractor, each corresponding to a different stable and coherent state. Which of these attractors governs a system's dynamics depends on the initial states or starting values of the system's evolution. The set of initial states that converges on each attractor represents the *basin of attraction* for that attractor. For a person or a group characterized by multiple attractors, then, the process in question can display different equilibrium tendencies. Within each basin, different initial states will follow a trajectory that eventually converges on the same value. Even a slight change in the system's initial state, however, will promote a large change in the system's trajectory if this change represents a state that falls just outside the original basin of attraction and within a basin for a different attractor.

Knowing the attractors and their associated basins of attractor for a social relationship is critical for understanding the effect that various events are likely to have on the relationship. The potential for multiple attractors in a system captures the intuition that people can have different (even mutually contradictory) goals, values, and patterns of social behavior. Consider, for example, a romantic couple that has two attractors: a strong attractor associated with positive feelings and a weaker attractor associated with negative feelings. Assume the attractor for positive feelings has a wider basin than does the attractor negative feelings. This suggests that the partners will evolve toward positive feelings if they begin an interaction within a broad range of affective states (e.g., neutral to very positive), but they may end up feeling negative about one another if they begin an interaction within a different (more restricted) range of affective states (e.g., mildly to highly negative). In other words, a broader range of initial states is likely to promote a communication trajectory that results in an exchange of warm sentiments as opposed to critical comments. On the other hand, if the couple routinely starts out with negative feelings, the negative attractor, despite having a narrow basin, may dictate the expression of feelings in the couple's interactions. By the same token, the couple may have a wider basin of attraction for negative feelings, so that anything short of a highly positive initial state could dissolve into a negatively toned exchange (Gottman, Swanson, & Swanson, 2002).

Additional Dynamical Patterns

Dynamical systems can also display more complex dynamical patterns as well (Guckenheimer & Holmes, 1983; Strogatz, 1994). These can include, for example, "limit cycles" that display repeated behaviors, such as the emotional cycling between depressive and manic states in bipolar disorders. They can also include "saddle surfaces" where behaviors of an individual or group first approach what might seem to be an attractor, but then rather than stabilize there, suddenly fly away into a different pattern. There are also situations where the behavior becomes so complex that it appears to be random even though the underlying low-level mechanisms that produce it are fully deterministic, a phenomenon that has been given the (perhaps paradoxical) name of "chaos." With a full understanding that dynamical systems can, and often do, display much more complex patterns of behavior, we will here concentrate on the concept of attractors, because the use of that concept alone already provides us a new perspective and novel insights into the understanding of social systems.

Latent Attractors

A system may have multiple attractors, but when the system is in the basin of one of them, the others may not be visible to observers—perhaps not even to the actors themselves. In fact, the existence of these potential states of the system might not even be suspected by observers or actors. These *latent attractors* nonetheless may be critical in the long run because they identify which states are possible for the system when conditions change. So although critical events and information might not be reflected in the observable state of the system, they may have the effect of creating a latent attractor representing a potential state that is currently invisible to all concerned. These changes may remain latent for extended periods of time, but can become manifest in response to external influences and events, even those that seem relatively minor. When the system falls within the basin of a latent attractor, the intrinsic dynamics of the system may show a dramatic (and unanticipated) change. This implies that the creation and destruction of latent attractors may be critical in shaping how the system will respond to events and conditions that have yet to take place. Such changes, however, are occluded from the vantage point of both participants and observers who know only the current state of the system.

Beyond characterizing a system's potential modes of behavior, recognition of latent attractors can prove critical in understanding (and implementing) plausible scenarios of change. Although movement between attractors can be rapid and abrupt, the change of attractors themselves is likely to be far slower and more gradual. So even though a specific policy or intervention does not produce a visible effect, this does not mean that it is doomed to futility. Rather, the policy or intervention may have the longer-term effect of creating or deepening a positive latent attractor for the system. By the same token, of course, the opposite scenario can be envisioned—negative events and interactions may fail to disrupt peaceful relations but gradually create and deepen a latent negative attractor characterized by mistrust, animosity, and violence. External forces and events, in other words, may not change the current state of the system but affect the range of possible states the system can adopt.

Attractor Dynamics in Perspective

Attractor dynamics are clearly relevant to a wide range of personal, interpersonal, and collective processes. Beyond capturing basic intuitions concerning these processes, framing them in terms of attractors allows for simple description of a system's dynamics. Instead of focusing on moment-to-moment changes in a system's state, one can describe the structure of attractors in the system and the patterns of transition between the attractors. Temporary and reversible changes in the state of the system reflect transitions between the system's existing attractors. Deep and lasting changes in the state of the system, meanwhile, correspond to changes in the structure of attractors.

Looking Ahead: Rethinking Conflict in Dynamical Terms

It is tempting to view conflict as an aberration of normal social psychological processes. Destructive conflicts that resist attempts at resolution seem especially pathological and thus unique in their underlying psychological and social processes. However, much the same could be said of blind obedience to authority, mob violence, and self-defeating behavior—or of heroism and passionate love, for that matter. Yet each of these domains of human behavior has been successfully framed in terms of basic processes that are manifest in far more banal realms of human experience. The dynamical perspective goes yet a step further in identifying common principles for seeming distinct topics. Indeed, the emphasis on basic dynamic processes provides a foundation for the subject matter of social psychology that unites this broad discipline with all the other disciplines of science. This chapter has shown in broad outline how social processes can be framed in terms of the dynamical perspective. The task taken up in the subsequent chapters is how to reframe conflict in terms of dynamical systems theory and, in so doing, provide coherence for the seemingly idiosyncratic manifestations of this feature of human nature.

References

Asch, S. E. (1946). Forming impressions of personalities. *Journal of Abnormal and Social Psychology, 41*, 258–290.

Asch, S. E. (1956). Studies of independence and conformity: A minority of one against a unanimous majority. *Psychological Monographs, 70* (Whole No. 416).

Axelrod, R. (1984). *The evolution of cooperation*. New York: Basic Books.

Axelrod, R., & Cohen, M. (2000). *Harnessing complexity: Organizational implications of a scientific frontier*. New York: Basic Books.

Axelrod, R., Riolo, R. L., & Cohen, M. (2002). Beyond geography: Cooperation with persistent links in the absence of clustered neighborhoods. *Personality and Social Psychology Review, 6*, 341–346.

Baron, R. M., Amazeen, P. G., & Beek, P. J. (1994). Local and global dynamics of social relations. In R. R. Vallacher & A. Nowak (Eds.), *Dynamical systems in social psychology* (pp. 111–138). San Diego: Academic.

Beek, P. J., & Hopkins, B. (1992). Four requirements for a dynamical systems approach to the development of social coordination. *Human Movement Science, 11*, 425–442.

Brehm, S. S., & Brehm, J. W. (1981). *Psychological reactance: A theory of freedom and control*. New York: Academic.

Carver, C. S., & Scheier, M. F. (1999). Themes and issues in the self-regulation of behavior. In R. S. Wyer Jr. (Ed.), *Advances in social cognition* (Vol. 12, pp. 1–105). Mahwah, NJ: Erlbaum.

Coleman, P. T., Vallacher, R. R., Nowak, A., & Bui-Wrzosinska, L. (2007). Intractable conflict as an attractor: A dynamical systems approach to conflict escalation and intractability. *American Behavioral Scientist, 50*, 1454–1475.

Condon, W. S., & Ogston, W. D. (1967). A segmentation of behavior. *Journal of Psychiatric Research, 5*, 221–235.

Cooley, C. H. (1902). *Human nature and the social order*. New York: Scribner.

Davis, B. P., & Knowles, E. S. (1999). A disrupt-then-reframe technique of social influence. *Journal of Personality and Social Psychology, 76*, 192–199.
Dittman, A. T., & Llewellyn, L. G. (1969). Body movement and speech rhythm in social conversation. *Journal of Personality and Social Psychology, 11*, 98–106.
Durkheim, E. (1938). *The rules of sociological method*. Chicago: University of Chicago Press.
Fiske, S. T., & Taylor, S. E. (1991). *Social cognition* (2nd ed.). New York: Random House.
Freud, S. (1920–1964). *Beyond the pleasure principle* (J. Strachey, trans.). New York: Norton.
Freud, S. (1937). *The ego and the mechanisms of defense*. London: Hogarth.
Gottman, J., Swanson, C., & Swanson, K. (2002). A general systems theory of marriage: Nonlinear difference equation modeling of marital interaction. *Personality and Social Psychology Review, 6*, 326–340.
Guastello, S. (1995). *Chaos, catastrophe, and human affairs: Application of nonlinear dynamics to work, organizations, and social evolution*. Mahwah, NJ: Lawrence Erlbaum Associates.
Guastello, S., Koopmans, M., & Pincus, D. (Eds.). (2009). *Chaos and complexity in psychology: The theory of nonlinear dynamical systems*. New York: Cambridge University Press.
Guastello, S. J., Pincus, D., & Gunderson, P. R. (2006). Electrodermal arousal between participants in a conversation: Nonlinear dynamics and linkage effects. *Nonlinear Dynamics, Psychology, and Life Sciences, 10*, 365–399.
Guckenheimer, J., & Holmes, P. (1983). *Nonlinear oscillations, dynamical systems, and bifurcations of vector fields*. New York: Springer.
Haken, H. (1978). *Synergetics*. Berlin: Springer.
Holland, J. H. (1995). *Emergence: From chaos to order*. Reading, MA: Addison-Wesley.
James, W. (1890). *Principles of psychology*. New York: Holt.
Johnson, S. (2001). *Emergence: The connected lives of ants, brains, cities, and software*. New York: Scribner.
Jones, E. E., & Gerard, H. B. (1967). *Foundations of social psychology*. New York: Wiley.
Kauffman, S. (1995). *At home in the universe*. New York: Oxford University Press.
Kelso, J. A. S. (1995). *Dynamic patterns: The self-organization of brain and behavior*. Cambridge, MA: MIT Press.
Krech, D., & Crutchfield, R. S. (1948). *Theory and problems of social psychology*. New York: McGraw-Hill.
Latané, B. (1981). The psychology of social impact. *American Psychologist, 36*, 343–356.
Latané, B., & Nowak, A. (1997). The causes of polarization and clustering in social groups. *Progress in Communication Sciences, 13*, 43–75.
Lewenstein, M., Nowak, A., & Latané, B. (1993). Statistical mechanics of social impact. *Physics Review A, 45*, 703–716.
Lewin, K. (1938). *The conceptual representation and measurement of psychological forces*. Durham, NC: Duke University Press.
Lewin, K. (1948a). *Field theory in social sciences*. New York: Harper & Brothers.
Lewin, K. (1948b). *Resolving social conflicts*. New York: Harper & Brothers.
McGrath, J. E., & Kelley, J. R. (1986). *Time and human interaction: Toward a psychology of time*. New York: Guilford.
Mead, G. H. (1934). *Mind, self, and society*. Chicago: University of Chicago Press.
Meinhardt, H. (1995). *The algorithmic beauty in sea shells*. Berlin: Springer.
Miller, K. D., Keller, J. B., & Stryler, M. P. (1989). Ocular dominance columns development: Analysis and simulation. *Science, 245*, 605–615.
Newtson, D. (1994). The perception and coupling of behavior waves. In R. R. Vallacher & A. Nowak (Eds.), *Dynamical systems in social psychology* (pp. 139–167). San Diego: Academic.
Nowak, A. (2004). Dynamical minimalism: Why less is more in psychology. *Personality and Social Psychology Review, 8*, 183–192.
Nowak, A., Szamrej, J., & Latané, B. (1990). From private attitude to public opinion: A dynamic theory of social impact. *Psychological Review, 97*, 362–376.

Nowak, A., & Vallacher, R. R. (1998). *Dynamical social psychology*. New York: Guilford.
Nowak, A., & Vallacher, R. R. (2001). Societal transition: Toward a dynamical model of social change. In W. Wosinska, R. B. Cialdini, D. W. Barrett, & J. Reykowski (Eds.), *The practice of social influence in multiple cultures* (pp. 151–171). Mahwah, NJ: Lawrence Erlbaum.
Nowak, A., Vallacher, R. R., Tesser, A., & Borkowski, W. (2000). Society of self: The emergence of collective properties in self-structure. *Psychological Review, 107*, 39–61.
Nowak, A., Vallacher, R. R., & Zochowski, M. (2005). The emergence of personality: Dynamic foundations of individual variation. *Developmental Review, 25*, 351–385.
Port, R. F., & Van Gelder, T. (1995). *Mind as motion: Explorations in the dynamics of cognition*. Cambridge, MA: MIT Press.
Read, S. J., & Miller, L. C. (Eds.). (1998). *Connectionist models of social reasoning and social behavior*. Mahwah, NJ: Lawrence Erlbaum Associates.
Schuster, H. G. (1984). *Deterministic chaos*. Vienna: Physik.
Smith, L. B., & Thelen, E. (1993). *A dynamic systems approach to development: Applications*. Cambridge, MA: MIT Press.
Strogatz, S. (1994). *Nonlinear dynamics and chaos*. Cambridge, MA: Perseus Books Group.
Strogatz, S. (2003). *Sync: The emerging science of spontaneous order*. New York: Hyperion Books.
Swann, W. B., Jr. (1990). To be adored or to be known? The interplay of self-enhancement and self-verification. In E. T. Higgins & R. M. Sorrentino (Eds.), *Handbook of motivation and cognition: Foundations of social behavior* (pp. 408–448). New York: Guilford.
Swann, W. B., Jr., Hixon, J. G., Stein-Seroussi, A., & Gilbert, D. (1990). The fleeting gleam of praise: Cognitive processes underlying behavioral reactions to self-relevant feedback. *Journal of Personality and Social Psychology, 59*, 17–26.
Tesser, A. (1978). Self-generated attitude change. In L. Berkowitz (Ed.), *Advances in experimental social psychology* (Vol. 11, pp. 85–117). New York: Academic.
Tesser, A., Martin, L., & Cornell, D. (1996). On the substitutability of self-protective mechanisms. In P. M. Gollwitzer & J. A. Bargh (Eds.), *The psychology of action* (pp. 48–68). New York: Guilford.
Thelen, E., & Smith, L. B. (Eds.). (1994). *A dynamic systems approach to the development of cognition and action*. Cambridge, MA: MIT Press/Bradford Books.
Tickle-Degnen, L., & Rosenthal, R. (1987). Group rapport and nonverbal behavior. *Review of Personality and Social Psychology, 9*, 113–136.
Turvey, M. T. (1990). Coordination. *American Psychologist, 4*, 938–953.
Vallacher, R. R., & Nowak, A. (Eds.). (1994a). *Dynamical systems in social psychology*. San Diego: Academic.
Vallacher, R. R., & Nowak, A. (1994b). The stream of social judgment. In R. R. Vallacher & A. Nowak (Eds.), *Dynamical systems in social psychology* (pp. 251–275). San Diego: Academic.
Vallacher, R. R., & Nowak, A. (1997). The emergence of dynamical social psychology. *Psychological Inquiry, 4*, 73–99.
Vallacher, R. R., & Nowak, A. (1999). The dynamics of self-regulation. In R. S. Wyer Jr. (Ed.), *Advances in self-regulation* (Vol. 12, pp. 241–259). Mahwah, NJ: Lawrence Erlbaum.
Vallacher, R. R., & Nowak, A. (2006). Coherence in human experience and psychological science. In P. A. M. Van Lange (Ed.), *Bridging social psychology: Benefits of transdisciplinary approaches* (pp. 77–82). Mahwah, NJ: Lawrence Erlbaum Associates Publishers.
Vallacher, R. R., & Nowak, A. (2007). Dynamical social psychology: Finding order in the flow of human experience. In A. W. Kruglanski & E. T. Higgins (Eds.), *Social psychology: Handbook of basic principles* (pp. 734–758). New York: Guilford.
Vallacher, R. R., Nowak, A., & Kaufman, J. (1994). Intrinsic dynamics of social judgment. *Journal of Personality and Social Psychology, 67*, 20–34.
Vallacher, R. R., Read, S. J., & Nowak, A. (2002). The dynamical perspective in social psychology. *Personality and Social Psychology Review, 6*, 264–388.

Vallacher, R. R., & Wegner, D. M. (1987). What do people think they're doing? Action identification and human behavior. *Psychological Review, 94*, 3–15.
Vallacher, R. R., & Wegner, D. M. (2012). Action identification theory. In P. A. M. Van Lange, A. W. Kruglanski, & E. T. Higgins (Eds.), *Handbook of theories of social psychology* (pp. 327–348). Thousand Oaks, CA: Sage.
van Geert, P. (1991). A dynamic systems model of cognitive and language growth. *Psychological Review, 98*, 3–53.
Waldrop, M. M. (1992). *Complexity: The emerging science at the edge of order and chaos*. New York: Schuster.
Wegner, D. M., & Vallacher, R. R. (1977). *Implicit psychology: An introduction to social cognition*. New York: Oxford University Press.
Weisbuch, G. (1992). *Complex system dynamics*. Redwood City, CA: Addison-Wesley.
Wolfram, S. (2002). *A new kind of science*. Champaign, IL: Wolfram Media.

Chapter 4
Patterns: Temporal Trajectories of Conflict

Despite the assumption of linear causality embedded in the last 400 years of science, it is almost impossible to predict specific outcomes in any nonlinear social system; their dynamics are too complex. Post-hoc determinations of causality can be made, but these are extremely difficult to ascertain a priori (Jones & Hughes, 2003). However, general *patterns* of thoughts, feelings, actions, and so on can be determined. For example, the exact behavior of a husband in a distressed couple at a given point in time cannot be predicted from knowing his attributes, initial states, and pre-existing conditions. But the pattern of his behavior in the marriage, over time, can be determined with knowledge of such parameters (see Gottman, Murray, Swanson, Tyson, & Swanson, 2002). A multi-dimensional plot of how the variables evolve in time—termed the *phase space*—provides a graphical image of the pattern of dynamics. A phase space may show evidence of both *attractor* regions (regions that are frequently occupied) and *repellor* regions (regions that are rarely occupied). Over time, in other words, certain behaviors become increasingly likely, habitual, and characteristic in a relationship (i.e., attractors), whereas other behaviors become increasingly unlikely and avoided (i.e., repellors). These regions can represent either positive or negative interactions (in terms of the effects on the relationship), and can be stable or unstable in a given relationship (Vallacher & Nowak, 1994a).

The pattern of activity of a system is determined by the underlying pattern of organization of the system. A system's *pattern of organization* is defined as "the configuration of relationships among the system's components that determines the system's essential characteristics" (Capra, 2002, p. 70). Thus, it is the pattern of organization of a system that will distinguish a normal cell from a cancerous one, a healthy family from a dysfunctional one, and a tractable conflict system from an intractable one.

Patterns of organization are believed to be strongly influenced by a set of conditions or choices made early on in the development of systems (Gersick, 1991; Gottman et al., 2002). Initial choices are thought to be particularly influential because they often characterize subsequent choices as appropriate or inappropriate, and become reinforced through positive feedback loops and through processes of

R.R. Vallacher et al., *Attracted to Conflict: Dynamic Foundations of Destructive Social Relations*, Peace Psychology Book Series, DOI 10.1007/978-3-642-35280-5_4,

self-organization. For instance, groups in conflict choose, to some extent, their group boundaries (who is in/who is out), allies, moral norms, leaders, and preferred methods of influence and conflict engagement. Their initial choices on these matters, whether automatic, reactive, or considered, will substantially affect the nature of the conflict dynamic that unfolds.

Consider, for example, Nelson Mandela and the African National Congress's choice to target infrastructure and not people in their militant campaign to end apartheid. This decision that had a tremendous impact on the development of that conflict and on the future of the country (Mandela, 1994). Similarly, research has indicated that mediators working with protracted conflicts face a basic set of choices upon entry to the system that are highly consequential (Coleman, Fisher-Yoshida, Stover, Hacking, & Bartoli, 2008). Decisions regarding who to work with (all stakeholders or select leaders), how to work (an open vs. clandestine process), how long to stay (demonstrating an enduring commitment vs. a perceived loss of neutrality), and what to work for (a comprehensive solution vs. a limited process of fair engagement) can be critical in shaping the patterns of conflict and peace that emerge over time.

From a dynamical point of view, it is important to distinguish between short-term and sustained changes in the dynamics of a conflict. Short-lived oscillations in conflict intensity or destructiveness are likely to be observed in any relationship characterized by attractor dynamics. Such short-term changes, however, can provide a misleading sense of the trajectory of the conflict. Although it is quite natural to experience hope and despair in response to changes in conflict intensity, they are not indicative of lasting changes as long as the underlying structure of the system remains stable. For lasting change to occur, the pattern of organization of the system must be modified in such a way that the attractor landscape is reconfigured.

In this chapter, we describe four dynamical models of basic conflict processes where short-term changes in outcomes (escalation, de-escalation, etc.) become comprehensible and predictable only when viewed in the context of their longer-term temporal patterns. The first model concerns cooperative and competitive conflict dynamics at the dyadic and intergroup levels, and examines the effects of differences in initial conditions and different strengths of mutual influence on those dynamics over time. The second model is a situated model of conflict in social relations (Coleman et al., 2012), which investigates the development of chronic versus adaptive conflict orientations in situations that evolve and change. The third model examines the importance of understanding how motivational orientations shift over time to promote optimal responses to conflict dynamics. And finally, we describe a dynamical model of conflict escalation that sheds light on linear versus non-linear (catastrophic) escalatory dynamics in conflict.

Non-linear Cooperation and Competition Dynamics

Dynamical systems can be used in different ways to understand scientific problems. In the physical sciences, equations are formulated to model systems and then solved to understand their dynamical behavior. Most of our applications of dynamical systems in social psychology have been to apply the qualitative lessons learned from that approach rather than to formulate and solve equations directly representing the phenomena studied. In this section, we will make use of that physical science approach. This model was inspired by Gottman et al. (2002), who took this approach to understand how wives and husbands interacted with each other. We begin by using key elements of their formulation but modify important features of their model so that it is applicable to understanding conflict. The results of the solution to the mathematical equations of the model give us interesting insights into how two groups might behave in a conflict. We have also used these insights to develop social psychology laboratory tests of some features of the model, which have also yielded interesting results.

It is important to note that our model is not intended to represent the many and diverse ways, on many different levels, that two groups interact in a conflict. The purpose of the model is to determine the logically necessary consequences of the social mechanisms that we have formulated in the equations and to discover what features of real conflicts might be attributable to the dynamics of such a simple model.

Gottman et al. (2002) modeled the state of each member of a married couple by difference equations that determined the present emotional valence of a member from (1) the emotional valence of that member a few moments ago, (2) the emotional valence when that member is alone, and (3) the influence of the present emotional valence of other member. In our model (Liebovitch et al., 2008; Liebovitch, Vallacher, & Michaels, 2010), we studied two interacting members, which could be two people or two groups of people or even two nations. For simplicity, we will use the word "group" to refer to each of these two members. Like Gottman et al., the emotional valence is the variable that represents the state of each group and we use a similar dependence of these variables on each group's previous state a few moments ago and its state when alone. Unlike Gottman, we chose influence functions that a number of authors had suggested should be appropriate to model a conflict (Deutsch, 1973, 2006; Pruitt, 2006) and we transformed the difference equations into differential equations so that they could be efficiently numerically integrated by standard software (ODE113 in Matlab).

The variables (x, y) in the model are the state (emotional valance) of each group. The equations of the model state that the rate of change of the state of each group (dx/dt, dy/dt) is proportional to the previous state of each group ($m_x x$, $m_y y$), the state of that group by itself (b_x, b_y) and the influence from the other group ($c_x f_x(y)$, $c_y f_y(x)$), namely,

$$dx/dt = m_x x + b_x + c_x f_x(y)$$

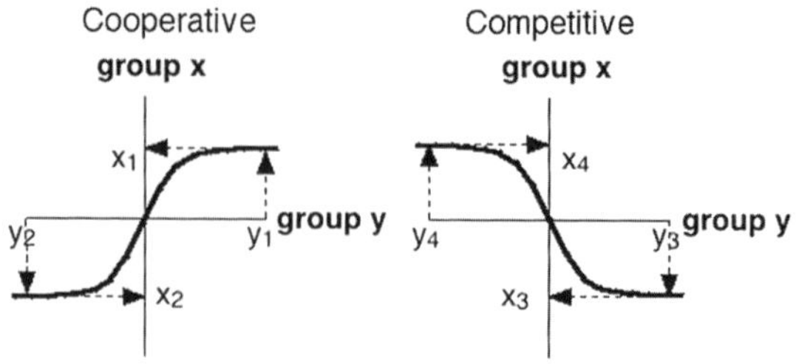

Fig. 4.1 The influence functions $f_x(y)$ that determine the response of group x to the state of group y when their relationship is cooperative (*left*) and competitive (*right*). The influence functions $f_y(x)$ of the response of group y to the state of group x would have a similar form

$$dy/dt = m_y y + b_x + c_x f_x(x)$$

Deutsch (2006) proposed that the interaction between two such groups could be characterized as either cooperation or competition: "To put it colloquially, if you're positively linked with another, then you sink or swim together; with negative linkage, if the other sinks, you swim, and if the other swims, you sink". We represented this interaction by the influence functions shown in Fig. 4.1, which are linear at low strengths of interaction and plateau at high strengths of interaction. In the cooperative interaction, each group reinforces the positive (or negative) state of the other group. In the competitive interaction, each group does the opposite of the other group.

We determined the dynamics of the model when (1) both groups are cooperative, (2) both groups are competitive, and (3) one group was cooperative and the other was competitive. We also determined how the dynamics depended on whether the influence between the groups was: (1) weak and (2) strong. The dynamical properties were found using a linear stability analysis (Gottman et al., 2002) to determine the attractors of these equations and their stability and we also numerically integrated the state (x, y) of each group to determine how that state evolves in time.

We found that this very simple dynamical model has some quite complex and interesting behaviors. That such complex behavior is contained in such simple assumptions is an example of "dynamical minimalism" (detailed in Chap. 3) and demonstrates the importance of that concept in understanding social psychological phenomena.

We found that this model had three important and interesting dynamical features:

1. The dynamics of both groups changes dramatically when the strength of the influence between them exceeds a certain threshold. Below that threshold, both groups evolve to a neutral state. Above that threshold (1) for the cooperative-cooperative case, both groups evolve to either a positive steady state or a negative steady state, (2) for the competitive-competitive case, one group

evolves to a positive steady state and the other to a negative steady state, and (3) for the cooperative-competitive case, the state of both groups varies periodically in time, slowly evolving to a neutral state. The groups evolve to steady states that depend sensitively on their initial state and their state when they are alone. For example, in the cooperative-cooperative case, if both groups start in a positive state or are in a positive state when alone, they will reinforce each other's state and both evolve toward a mutual positive state. However, if one (or both) starts in a sufficient negative state or a sufficiently negative state when alone, then they will both evolve toward a negative steady state.

2. Since the dynamics depends on the actions of both groups, one group alone by changing its actions can cause both groups to evolve toward a new and different steady state. This is quite a novel result. It had been thought that conflict resolution depends on first one group changing its strategy, which then induces a similar change of strategy in the other group. Our finding here is that since the dynamics depends on the actions of both groups, one group alone through a judicious choice in changing its own actions can change the dynamics of both groups. This result is present in two different scenarios in the model. First, the difficult steady situation where one group is negative and the other is positive in the competitive-competitive case can be changed to a much improved situation where both groups are neutral by the one negative group switching its strategy to cooperative. This occurs because the cooperative-competitive case results in both groups evolving toward neutral states. Second, carefully timed switches of actions can lead both groups to evolve to different steady states. As already noted, the competitive-competitive case results in the difficult steady situation where one group is negative and the other is positive. If the negative group switches to cooperative, the dynamics now changes so that the state of both groups varies periodically in time. If that same group now switches back to being competitive at a time when it is more positive than the other group, both groups will evolve to states that are reverse of their initial interaction. That is, the group that initially reached the positive steady state will now be at a negative steady state and vice versa. (In these dynamical models, the steady state of both groups depends on their dynamics—that is, how their parameters vary in time. Thus the groups in these dynamical models can reach steady states that would not predicted by the static energy assignments of game theory models.) Thus, this surprisingly simple model has yielded some very interesting results.

We next sought to test whether some of the dynamical behavior predicted by the model could be observed in an experimental social psychology setting. One feature of the model is that it predicts that periodic oscillations in the state of both groups should be present when one group is cooperative and the other is competitive, but not when both groups are cooperative or both groups are competitive. We developed an experimental situation where the attitude similarity and valence of participants was measured with interactions with confederates (Michaels, 2012). The participants and confederates alternated in asking and answering questions with each other. The responses of the confederates were either neutral, consistently cooperative, consistently competitive, switched from cooperative to competitive at the midpoint of the session, or switched from competitive to cooperative at the midpoint of the session.

To vary the strength of the interaction between the participant and confederate, a pretest was used to determine the "hot button" topics of the participant, which were then either used (strong influence) or not used (weak influence) in the study. It was found that when the confederates switched from cooperative to competitive behavior, the participants exhibited significantly greater attitude change over the course of the interaction than did participants in the all the other groups. This study provides some evidence that switching from cooperation to competition does induce considerable changes. This study did not have the time resolution to detect the oscillations predicted by the model.

In summary, it is possible to use the paradigm of the physical sciences to investigate systems in social psychology in general and in conflicts in particular. That is, we can represent social mechanisms as the terms in rigorous mathematical equations, use mathematical and computational methods to solve those equations to determine the logically necessary consequences of those assumptions, and test those results with empirical experiments. The complex dynamical behavior of this very simple model and the dependence on its parameters may provide insight into some of the behaviors observed in conflicts. Such a model does not, and is not intended, to reproduce all the multi-faceted and multi-level interactions present in actual conflicts, but it may prove helpful in understanding some dynamical features of such conflicts.

Chronicity and Adaptivity in Situated Conflict Dynamics

How situations shape people and people in turn shape situations dynamically has been perhaps *the* topic of social science since first debated by Aristotle and Plato and later studied by the likes of Darwin and Freud (see also Mischel, 1977). Research on conflict has likewise investigated how both social conditions and individual differences affect conflict dynamics and outcomes (see De Dreu, 2010; Deutsch, 2002; Deutsch, Coleman, & Marcus, 2006; Kriesberg, 2003; Pruitt & Kim, 2004 for comprehensive summaries). However, most of these studies have tended to focus on separate aspects or dimensions of conflict, such as different motives, styles, conditions, degrees of dependence, or levels of power (Coleman et al., 2012), and have tended to neglect the interactive dynamics of person-situation variables in conflict. This has left our understanding of person-situation conflict dynamics in a rather piecemeal state.

Dimensions of Social Relationships

However, a recent model offers the promise of an integrative framework for the study of person × situation conflict dynamics over time (Coleman, Vallacher, Bartoli, Nowak, & Bui-Wrzosinska, 2011; Coleman et al., 2012). Rather than isolating distinct variables and investigating their separate effects, it builds on prior research identifying the fundamental features of social relations (Deutsch, 2007, April, in press; Wish, Deutsch, & Kaplan, 1976). In particular, the model holds that three basic aspects of social relations are highly relevant to conflict in relationships. First, social relations can have different *degrees of goal interdependence*. To what extent does a person see his or her goals linked to those of the other party? The greater the degree of such interdependence, the more important the relationship is for the parties. Second, social relations can have different *forms of goal interdependence*. What is the ratio of cooperative to competitive goal interdependence in the relationship? This aspect captures the person's perception that the other party is either for or against him or her, or perhaps has a mix of cooperative and competitive tendencies. Third, social relations can be characterized in terms of the *relative power* of the parties. Does the person see him or herself having greater, less, or the same power (resources, information, status, etc.) *vis a vis* the other party in the relationship?

These three aspects are independent and thus can combine in various ways, with each combination representing a unique representation of a relationship (see Fig. 4.2). Thus, one person might see his or her relationship with someone else as cooperative, with relatively equal power, but with little goal interdependence (i.e., low relational importance), whereas another person might see his or her relationship with someone as competitive, with relatively equal power, and with high goal interdependence (high relational importance). A person's representation on each dimension can reflect relatively objective considerations (e.g., unambiguous differences in resources or status, signifying a power differential), but it can also reflect individual differences in chronic psychological orientations to conflict (e.g., a heightened sensitivity to the prospect of competitive interdependence in social relations).

A person's unique representation within this three-dimensional framework can also change and evolve over time, with important implications for the quality of the relationship as it unfolds. Indeed, the model attaches considerable importance to the temporal stability versus variability of people's view of the relationship on each dimension. Whether a person's representation remains stable or undergoes noteworthy change, in turn, can reflect objective circumstances (e.g., a change in relative power) or the salience of his or her chronic orientation (e.g., greater or less sensitivity to power differential as a relationship evolves) (Deutsch, 1982; Kelley, 1997; Mischel, 1977). The framework, then, is useful not only for characterizing a person's representation with respect to essential features of social relations, but also for tracking the evolution of the person's representation in response to objective factors and subjective concerns.

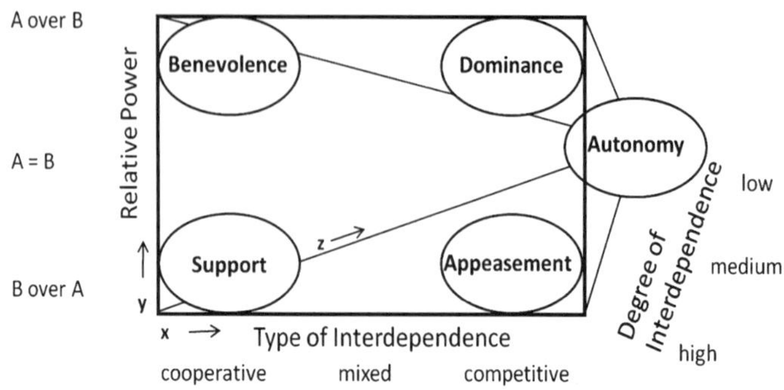

Fig. 4.2 The situated model of conflict in social relations (Coleman et al., 2012)

The Model in Action

A central proposition of the model is that very different psychological orientations to conflict are induced depending on the person's specific representation within the three-dimensional framework. Psychological orientations are a complex of cognitive, motivational, moral, and action orientations to a given situation that serve to guide one's behaviors and responses (Deutsch, 2007, in press). Special consideration is given to five specific combinations of a person's representation on the three dimensions (see Fig. 4.2) because each induces distinct perceptions, values, emotions, and behaviors relevant to conflict (see Deutsch, 1982, 1985, 2007, in press; Kelley, 1997; Van Lange et al., 1997). In particular, these configurations are hypothesized to evoke very different conflict orientations, which partially determine (1) how conflicts are perceived (as mutual problems or win-lose challenges), (2) how it feels to be in the situation (relatively comfortable versus anxiety-provoking), (3) what is likely to be valued in the situation (solving problems and sharing benefits with other parties versus conquering them), and (4) how to best go about responding to the conflict and obtaining these values and goals (through respectful dialogue and problem-solving versus forceful domination or submission to power). The five psychological orientations are labeled (1) *benevolence* (afforded by high power, cooperative, high interdependence), (2) *dominance* (afforded by high power, competitive, high interdependence), (3) *support* (afforded by low power, cooperative, high interdependence), (4) *appeasement* (afforded by low power, competitive, high interdependence), and (5) autonomy (afforded by low degrees of interdependence) (see Fig. 4.2).

Research testing the model (Coleman et al., 2010, 2011, in press; Kugler & Coleman, 2010) found that when participants were presented with the same conflict (in terms of the goals and issues), but were situated in different regions of the stimulus field, they described markedly different experiences—perceptions, emotions, values, and behavioral intentions. For instance, when faced with a 'high-power, cooperative, high interdependence'-scenario, participants described a more active-cooperative orientation to conflict than they did for most of the other regions—where participants said they valued taking responsibility for the problem, listening to the other, and expressed genuine concern for their low-power counterpart (benevolence). In contrast, 'high-power, competitive, high-interdependence' scenarios *for the same conflict issues* were found to induce a more threatening and confrontational approach to the other party, less concern for the other, and heightened concerns for their own power (dominance). Similar congruent situation-orientation findings were identified for the support, appeasement, and autonomy regions of the field.

The model also proposes that when parties become situated in a particular region of the stimulus field for extended periods of time (e.g., stuck in low-power in a competitive relationship), they tend to develop a strong orientation for that region, which can become *chronic*. Research on high-power/low-power party dynamics provides evidence of such chronicity in conflict orientations. Magee and Galinsky (2008), for example, cite research detailing the psycho-social transformations that take place in organizations that promote a chronic dominance orientation among powerful parties, buttressed by their ongoing attempts to retain and acquire power. Case studies of state-level international trade negotiations also provide support for the idea that high-power parties become very comfortable with dominance orientations, and find it difficult to employ other strategies when power shifts or conditions change (Zartman & Rubin, 2002). This research also suggests that low-power parties tend to become accustomed to their complementary role and skilled in performing it.

The situated model further predicts that conflictful relations between parties that remain quasi-stable over an extended period of time will lead to the *internalization of conflict orientations for the relationship*. Zartman and Rubin (2002) found evidence of this type of coordination dynamic between high- and low-power negotiators in their analysis of international conflicts. When high-low power relationships between parties were well-established through an on-going relationship (e.g., U.S. and Mexico), the parties tended to know their relative-power roles and enact them in a complementary manner. Even if important changes in the basic parameters occur (such as power shifts), the strength of the relational orientation may maintain the original dynamic, or at least delay significant changes in the dynamic. When this occurs, people may find it emotionally distressing when situations arise which require that they adopt a different approach in the relationship (McClelland, 1975; Rusbult &Van Lange, 2003).

Although specific psychological orientations to conflict may be useful and more fitting in particular situations, problems typically arise for people when they become fixated on any one orientation, or when an individual's chronic orientation(s) fits poorly with the specific realities or demands of situations. Each of the five orientations outlined in the model has its particular utilities and benefits, costs and consequences, depending on the psychological makeup of disputants, the orientations of other parties, and the nature of the situations faced. In fact, all of the orientations, when chronic, have their associated pathologies (see Coleman et al., 2012). Ultimately, the situated model suggests that more *adaptive* orientations to conflict—those that allow for the use of different orientations and strategies in order to satisfy goals in a manner not incongruent with the demands of the situations encountered—should lead to greater efficacy and satisfaction with conflict processes and outcomes over time.

Research has provided support for this idea as well. Case-based research on interstate negotiations found that parties tended to be more effective in negotiations to the extent that they were able to adjust their orientations and behavior to the relative (and relevant) power of the other side (Zartman & Rubin, 2002). In a correlational study (Coleman, Mitchinson, & Kugler, 2009), investigators found that more adaptive individuals (individuals who reported utility in employing all five orientations when necessary) had greater levels of satisfaction with conflicts in general, learned more from conflicts, and were better able to focus on both long-term and short-term goals than were less-adaptive individuals. Using a new measure for conflict adaptivity in the workplace that builds directly on the described model, Coleman and Kugler (2011) found that more adaptive managers are also more satisfied with conflict processes at their own workplace and report higher job-related well-being in general. A third study, conducted through the collection of critical-incidents of actual conflicts in participants' lives, found that people who were able to employ orientations and behaviors that were not incongruent with the situation (not ill-fitting) expressed greater satisfaction with their own behavior, as well as with the processes, outcomes, and relationships involved in those conflicts (Coleman & Kugler, 2011).

To summarize, research on the situated of social conflict sheds new light on some of the more fundamental patterns and dynamics involved in social conflict, such as the way in which the three parameters of the model interact to affect conflict orientations, the importance of adaptivity to long-term conflict management and satisfaction, and how protracted conflicts can lead to the internalization of relational dynamics which resist change. The value of the model, then, is not the identification of new factors and variables, but rather in delineating how a minimal number of these factors—those deemed the most essential—is sufficient to capture the complexity of conflict in a wide range of interpersonal and inter-group relations. It also highlights the importance of *movement* in constructive conflict resolution. Instead of emphasizing a set of predispositions or conditions in positive conflict processes, the model stresses the necessity of adaptation; conflicts can be constructively managed when the disputants are able to move between different orientations, strategies, and tactics as evolving situations require. As such, the model provides a solid step forward in our understanding of conflict dynamics in the broader context of social relations. In addition, the development of the situated model provides us with an important platform to begin to investigate the dynamical properties of the model, such as the role by initial conditions, attractors, nonlinearity and catastrophic transitions in how patterns of social conflict unfold.

Regulatory Focus, Attractor Dynamics and Optimality in Conflict

Although the motives for any conflict can be numerous, when viewed in terms of basic dynamics they can be viewed as reflecting one of two basic orientations toward experience: achieving gain (rewarding outcomes) versus avoiding loss (aversive outcomes). In contemporary social psychology, this distinction has been reframed in terms of regulatory focus (Higgins, 1997, 2012), which distinguishes between two primary modes of experience: promotion and prevention.

Promotion Versus Prevention Orientations Toward Conflict

When in a promotion focus, people are especially sensitive to experiences that enhance positive outcomes, whether material (e.g., attaining money or resources) or psychological (e.g., gaining power or enhancing self-esteem) in nature. In a prevention focus, people are especially sensitive to experiences that threaten to undermine positive outcomes or increase the risk of negative outcomes. These modes do not represent endpoints on a single scale, but rather two distinct ways of scaling experience. Thus, the same state of affairs can be valued to the extent that it promotes an advance in one's well-being (promotion mode) or to the extent that it prevents an erosion of one's current level of well being or reduces the intensity of negative outcomes (prevention mode). For example, a conflict in which nothing bad happens might be viewed as unsatisfying from a promotion perspective, but as quite satisfying from a prevention perspective. As such, promotion and prevention can be conceptualized as distinct attractors for psychological dynamics.

With respect to conflict, this perspective suggests that a conflict can develop and resist resolution for two very different reasons: to enhance personal or collective outcomes or to avoid deterioration in an existing state of affairs. Thus, one party to a conflict may view the conflict as an opportunity to increase its access to resources or to enhance its status, whereas the other party may view the conflict as necessary to avoid a loss of resources or an erosion of status. Emphasizing the gain to be had from conflict resolution may be persuasive for one party, but emphasizing the cessation of threat and resource depletion may have greater resonance for the other party. The salience of these orientations—and the associated relevance of "fit" in conflict resolution (whether one's orientation is consistent with a particular conflict management process; see Higgins, 2000)—is likely to be observed when there is an increase in intensity of conflict and a corresponding collapse of complexity in the parties' perception of issues and of one another. Here, the parties more chronic orientations will likely take hold and drive their perceptions and responses to the conflict.

Research to date on regulatory focus has emphasized individual differences in promotion versus prevention. Generally speaking, this research has shown that compared to prevention-focused individuals, promotion-focused people achieve superior outcomes in negotiations, prefer gain-maximizing strategies, and are more creative and open to change, but are also more likely to retaliate against transgressions (e.g., Appelt & Higgins, 2010; De Cremer, Brebels, & Sedikides, 2008; Galinsky et al., 2005; Ten Velden et al., 2009).

We have found, however, that there are specifiable links between the changing circumstances associated with conflict as it unfolds and variance in the relative salience of promotion and prevention modes. In unfolding processes of social conflict, these modes can operate in combination or in an iterative fashion as conditions change and affect the relative salience of these orientations. These situational changes may then act to override disputant's predisposed tendencies for prevention or promotion. This is particularly likely in mixed-motive conflicts (where the parties have both common and competing goals operating; see Deutsch, 1973) and in conflicts that escalate over time and thus promote substantive changes in the salient motives (Fisher & Keashly, 1990; Pruitt, Kim, & Rubin, 2004). Under these conditions, both loss and gain are important.

Preliminary Research on Regulatory Focus in Conflict

We tested this general idea in three studies that investigated the relationship between different modes of regulatory focus and social conflict over time (Coleman, Kugler, & Vallacher, 2013). The studies extended prior research by examining the combined effects of prevention and promotion orientations on experiences of conflict, and by employing a method for studying the ratios of promotion-to-prevention orientation.

The first study was conducted as an on-line survey to test correlations between trait measures of chronic prevention or promotion regulatory focus and distinct emotional, value, and behavioral intention syndromes across three conflict scenarios. In line with the predictions of regulatory focus theory (Higgins, 1997), and previous findings on regulatory focus and conflict (e.g., Appelt & Higgins, 2010; Galinsky et al., 2005; Ten Velden et al., 2009), we found support for our hypothesis that people who undertake and evaluate actions in their daily life with a promotion orientation tend to approach conflict in the same way, focusing on the opportunity for positive outcomes, being sensitive to hits and errors of omission, and experiencing correspondingly positive emotions. In contrast, we found that people who undertake and evaluate daily actions with a chronic prevention orientation have heightened sensitivity in conflict for negative outcomes, for correct rejections, for errors of commission, choose an avoidance rather than approach strategy, and experience more negative emotions. However, should one look to prevent problems, seek ideal solutions, or both when facing social conflict?

The second study addressed the question that given the mixed-motive nature of most disputes—which present both opportunities and threats and elicit both hopes and fears—what is an optimal orientation to conflict? We tested this experimentally by framing a conflict scenario as either prevention-focused, promotion-focused, or as a combination of both. As hypothesized, a combined prevention and promotion orientation promoted the highest satisfaction with conflict outcomes and goal attainment. However, framing the conflict as promotion focused promoted the highest levels of satisfaction with the process of conflict engagement and with the interpersonal relations inherent in this exchange.

Ultimately, we sought to investigate how regulatory focus orientations operate over time to affect conflict experiences and outcomes. This study employed the mouse paradigm (Vallacher & Nowak, 1994b; Vallacher, Nowak, & Kaufman, 1994), a method for coding temporal data, to explore relative time spent in a promotion versus prevention orientation while recalling a critical incident of conflict. This involved participant self-coding of stream-of-thought accounts of actual conflicts to investigate how different ratios of prevention-to-promotion orientation were associated with differences in satisfaction with conflict processes, relationships and outcomes. Again, this study found that a balanced focus on both prevention and promotion was associated with more optimal outcome satisfaction in conflicts than was either prevention or promotion alone, whereas a higher ratio of promotion-to-prevention focus was associated with higher process and relationship satisfaction.

Considered together, these studies suggest that it is important to identity the criteria of "success" when deciding how best to approach conflict transformation. If one's primary concern is maximizing satisfaction with the *process* of conflict engagement, emphasizing a promotion orientation in both parties is advisable. However, the results of our research to date suggest that a balance between promotion and prevention focus may be optimal for attaining one's *goals and outcomes* in evolving social conflicts. In a broader sense, our results highlight the importance of better understanding the concept of optimality and its role in the moment-to-moment and long-term dynamics of conflict and conflict resolution. The construct of *optimality* has been investigated in terms of decision making (Davis-Stober, Dana, & Budescu 2010; Nelson, 2009; Pollock, 2006; Tamura, Kameda, & Fukano, 2006), personality (Sheldon, 2007), and emotions (Gottman et al., 2002; Kristjánsson, 2009; Losada & Heaphy, 2004). However, it has rarely been explored in the context of motivation or conflict management (see Callanan, Benzing, & Perri, 2006; Euwema, Van de Vliert, & Bakker, 2003; Van de Vliert, Euwema, & Huismans, 1995). Rather than claiming that any one tendency (e.g., promotion, cooperativeness, high self-esteem, open-mindedness, sensitivity) is the "best"—an idea which is implicit in many theories—we suggest that a balance between tendencies may often be the most desirable and effective. The task for further research is to identify the optimal balance between tendencies as a function of various factors. With respect to regulatory focus, in particular, research in progress is investigating the optimal balance between promotion and prevention for conflicts that vary with respect to parameters like power differentials, sources of conflict (e.g., resources vs. ideology), and types of interdependence (e.g., confined vs. voluntary, etc.).

The Dynamics of Conflict Escalation

Conflict escalation, defined as intensification of conflict over time, is usually viewed as negative, highly destructive, and difficult to manage. This is particularly true when people engage in relationship conflict (Canary, Cupach, & Messman, 1995; Deutsch, 1977; Jehn, 1997), competitive cycles of escalation (Pruitt & Kim, 2004), enduring rivalries (Goertz & Diehl, 1992, 1993), or when escalation reaches a point at which relations turn into deeply malignant processes (Deutsch, 1973) and conflict becomes fundamentally intractable (Bar-Tal, 2000; Coleman, 2003; Kriesberg, 2005). On the other hand, in certain cases increasing conflict to an optimal level of intensity may prove useful for learning (Johnson, Johnson, & Smith, 2000), team creativity and performance (De Dreu, 1997; Jehn, 1995, 1997), satisfactory close relationships (Gottman et al., 2002; Pruitt & Kim, 2004), or, at the macro level, constructive maintenance of peace and democracy (Reykowski, 2006; Reilly, 2001).

So why do some conflicts escalate to destructive levels of intensity, while others prove necessary for overcoming stagnation and fostering constructive change? How can we better understand the conditions and processes that move people's relations from relatively mild and constructive disagreement to coercive, destructive interactions? To answer these questions, it is critical to look at the process of escalation as it unfolds over time.

Linear Versus Non-Linear Scenarios of Escalation

Attractor dynamics provide a particularly fruitful way to characterize escalation trajectories. Peaceful relations correspond to the existence of a strong attractor characterized by positive interactions. If an external event moves the system out of its attractor (e.g., a momentary increase in hostility), the system will shortly return to its attractor (e.g., the parties will resolve the issue). With increased provocation (e.g., sustained hostility by one party), the positive attractor will be weakened or replaced at some point by an attractor corresponding to negativity. There are two different routes by which this may happen (Thom, 1975). In a *linear scenario*, increased provocation on one side will result in a gradual shifting of the positive attractor to increasingly negative values (e.g., from friendly to neutral to unfriendly to hostile). Although the value of the attractor changes, it represents the only attractor for the system. In a *non-linear scenario*, the initial attractor for peaceful interactions doesn't change its value despite increasing provocation, but becomes progressively weaker. At some threshold value of provocation, a second latent attractor at high values of negativity is created or activated, polarizing the scope of interactions. With yet further provocation, the positive attractor weakens and the negative attractor strengthens. At some point, the positive attractor loses its stability and the relationship abruptly moves to the extreme values defined by the negative attractor, which then governs the dynamics of the relationship.

De-escalation mirrors the escalation scenarios. In the linear case, the attractor moves incrementally to positive values. In the nonlinear case, the progression of reconciliatory actions weakens the negative attractor and reinstates the positive (latent) attractor. At some point, the system abruptly switches from values defined by negative interaction to values defined by positive interaction—in effect, the positive attractor becomes manifest and the negative attractor becomes latent. This switch is likely to occur at higher positive values of interaction than the values at which the system switched from negative to positive. This phenomenon, in which the system displays a tendency to remain at its current attractor, is termed *hysteresis* in catastrophe theory (Thom, 1975) and represents a signature of nonlinearity. Even if the interaction between parties changes to quite positive values, the presence of a latent negative attractor is indicative of a tendency of a system to return to high negativity in response to even slight provocations.

Factors Influencing Linear Versus Nonlinear Escalation Scenarios

Whether the escalation of a conflict in response to provocation follows the linear or nonlinear scenario does *not* depend on the type of conflict (e.g., interpersonal or international, resource- or identity-based), the type of provocation (e.g., physical vs. verbal), or even the relative power of the parties (e.g., equal vs. asymmetric) to the conflict. These factors are certainly relevant to how a conflict unfolds over time, but they do not play a prominent role in shaping the temporal pattern of the conflict. So what does determine whether a linear or a nonlinear trajectory is observed in conflicts characterized by increasing provocations, whether by one party or both?

From the perspective of dynamical systems, the *complexity* of a system is crucial in determining how it behaves in response to perturbing influences. Complexity refers to the differentiation or functional independence of a system's lower-level elements. Two systems may have the same number of elements, yet differ a great deal in the degree to which the elements function independently or instead demonstrate mutual influence by virtue of feedback loops among them. If the elements function independently, so that a change in one element has little or no impact on the state of other elements, the system has high complexity—or many *degrees of freedom* (cf. Holland, 1995a, 1995b). If, on the other hand, the elements exert strong mutual influence, so that a change in one element promotes change in other elements, the system has low complexity—or few degrees of freedom.

In a system characterized by high complexity, an external influence on one element is likely to be confined and thus may not change the state of other elements. In a system characterized by low complexity, however, anything that changes the state of one element is likely to have ripple effects throughout the entire system, as the feedback loops among the elements are engaged. Initially, however, this impact may not be apparent because the connections among affected elements are likely to absorb the impact. In effect, the change in state of one element is corrected (by feedback loops) by the current state of many elements. Once a threshold of impact is reached, however, the states of all the connected elements may suddenly change in tandem, promoting a significant (catastrophic) change in the system's overall state. In effect, a change in one element has a domino effect, producing a cascade of changes throughout the system.

The distinction between low and high complexity has direct implications for the likelihood of linear versus nonlinear escalation scenarios. When a provocation is experienced in a system of high complexity, its impact on the system's behavior will be confined to relevant portions of the system and thus is unlikely to change the global properties or behavior of the system. In marked contrast, the same provocation experienced in a system of low complexity is far less likely to have limited impact over time. Even if the provocation concerns a single element, the links between that element and the other system elements create the potential for widespread effects on the system, perhaps changing its global behavior in a qualitative manner. In the former case (high complexity), the system will change in a proportional (linear) manner in response to the provocation. In the latter case, the system will change in a disproportionate (nonlinear) manner in response to the provocation. The change may not be apparent immediately—in fact, the system may appear highly resistant to perturbation—but at some threshold of perturbation intensity, the system may display catastrophic change.

The contrast between these two generic scenarios—linear change in a system with high complexity, nonlinear change in a system with low complexity—is depicted in Fig. 4.3. The first Fig. 4.3a represents the linear scenario, characteristic of a social system with high complexity (many degrees of freedom). As the level of provocation by one party increases, the other party responds in a proportional manner, matching the degree of provocation in a tit-for-tat manner. The second Fig. 4.3b represents the nonlinear scenario, characteristic of a social system with low complexity (few degrees of freedom). As the level of provocation by one party increases, the other party shows little or retaliation at first, but then displays a disproportionate response in line with catastrophe models (cf. Thom, 1975). The same pattern is observed when the level of provocation is reversed, with strong retaliation giving way in a catastrophic manner to relatively mild retaliation. Note the hysteresis, however, in Fig. 4.3b. The point at which increasing provocation produces a nonlinear increase in retaliatory response is not the same point at which decreasing provocation produces a nonlinear decrease in retaliatory response. As noted earlier, hysteresis is a signature phenomenon of nonlinear dynamical systems.

Preliminary Research on Linear Versus Nonlinear Escalation

The difference between linear and nonlinear escalation is relevant to systems at different levels of psychological and social reality. Whether the conflict is between two people, two groups, or two nations, the degree of complexity in each party's

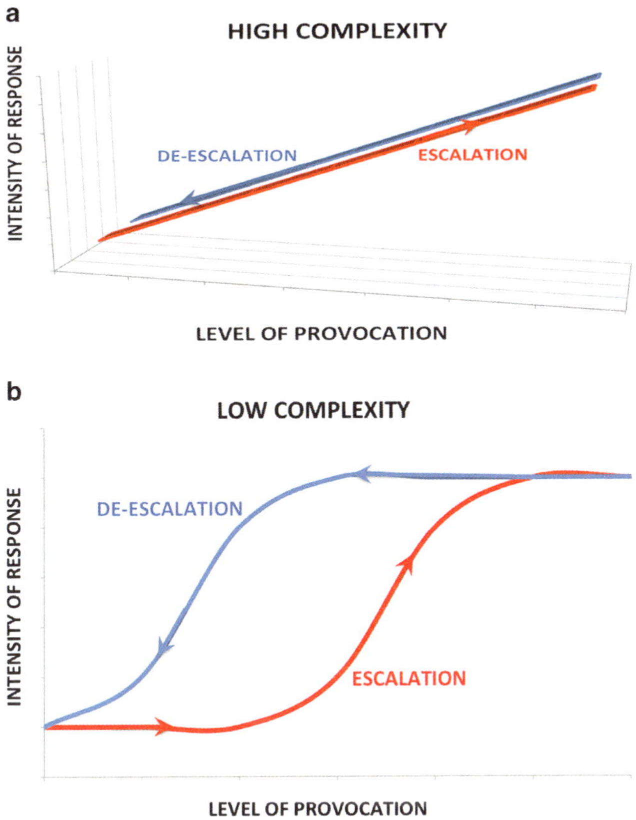

Fig. 4.3 Conflict escalation trajectories for high versus low complexity structures

representation of the relationship will dictate whether a conflict escalates in a proportional or catastrophic manner. We have tested this idea in two lines of laboratory research—one focusing on the complexity of an interpersonal relationship, the other focusing on factors that enhance versus reduce the level of complexity in the representation of a conflict.

Escalation of Conflict in Interpersonal Relationships

Bui-Wrzosinska (2005) investigated the role of complexity in escalation of conflict by investigating the temporal pattern of hostility in social relationships that were presumed to differ in their degree of relative complexity. In a simulated school setting, participants were asked to imagine working with a fellow student on a common project. At each succeeding stage in their collaboration, the peer acted in an increasingly provocative manner toward the participant (e.g., mild and private criticism initially, extreme and public insults at a later stage). Participants were provided a list of behaviors that varied in hostility (moderate to very destructive), and they were asked to indicate which of these behaviors were appropriate responses to the peer's provocation.

To test the hypothesis that system complexity dictates whether a linear versus a nonlinear scenario characterizes the escalation of retaliation, Bui-Wrzosinska asked the participants to imagine that the peer either a personal friend (close) or a casual acquaintance (distant). Research on social cognition has shown that cognitive representations of acquaintances tend to differentiated, with the various dimensions of judgment functionally independently, thus making an extreme global assessment unlikely (cf. Fiske & Taylor, 1991; Wegner & Vallacher, 1977). For example, a person may view an acquaintance in positive terms with respect to fairness, in negative terms with respect to conscientiousness, and in neutral terms with respect to energy level. In a distant relationship characterized by complexity, then, the overall judgment is likely to have relatively moderate valence, reflecting a linear combination of the values associated with each of the dimensions (cf. Anderson, 1981). Change in overall judgment should be proportional to the changes in the values associated with relevant dimensions. Hence, under increasing provocation by the peer, participants were expected to choose behavioral responses of gradually increasing hostility.

In a close relationship, however, each person is likely to have a relatively global judgment of the other person, with different aspects of the person seen as highly correlated (Asch, 1946; Rosenberg & Sedlak, 1972; Wegner & Vallacher, 1977). If a person is judged to be fair, for example, he or she is also likely to be judged positively on other dimensions (sincerity, energy, etc.). In dynamical terms, this "halo effect" (Thorndike, 1920) means that the specific dimensions of judgment are linked by positive feedback loops, thus reflecting a low complexity representation. This renders the overall judgment difficult to change because each element of the judgment is protected against the influence of new information that might contradict the current valence of that element (cf. Latané & Nowak, 1994; Nowak, Vallacher, Tesser, & Borkowski, 2000). Beyond a certain value, however, new information may transform the valence of a highly important or central judgment dimension (e.g., from fair to unfair). Because of the positive feedback loops between this dimension and the other dimensions, this change promotes a wholesale change in the global judgment. In effect, the global judgment characterizing a low complexity (e.g., close) relationship resists change until a critical threshold is reached, at which point the judgment displays a catastrophic change in valence.

Results confirmed this hypothesis. When participants experienced increasingly provocative behavior from the close personal friend, they responded in a relatively mild fashion until a critical threshold of antagonism was reached, at which point they began to respond with highly confrontational behaviors. The transition from one type of response (mild) to the other (hostile) was abrupt and did not involve a transition through intermediate levels. However, when the relationship with the peer was distant (characterized by weak linkage among the various judgment dimensions), participants responded in a proportional manner to antagonistic actions, with escalation and de-escalation of conflict intensity following a linear scenario. Intermediate degrees of antagonism prompted retaliatory responses of intermediate intensity.

Another group of participants experienced a de-escalating pattern of provocation, starting with high provocation and progressing through conciliatory gestures. The results showed that the change in de-escalation was non-linear for close relations and a linear for distant relations. Moreover, the close group maintained the hostile response tendency until the conciliation efforts by the peer were clearly pronounced, whereas the distant group reacted sooner to such gestures. This evidence of hysteresis supports the contention that conflict in close relationships represents a nonlinear dynamical system.

Need for Closure and Conflict Escalation

People differ in their concern for achieving a unified and stable judgment of events and people, and the salience of this concern is inversely related to tolerance of ambiguity, complexity, nuance, and uncertainty. In dynamical terms, this judgment tendency can be seen as a manifestation of the *press for coherence* (cf. Coleman, Bui-Wrzosinska, Vallacher, & Nowak, 2006; Coleman, Vallacher, Nowak, & Bui-Wrzosinska, 2007) that promotes progressive integration of system elements into a higher order structure. Individual differences in this tendency have been the focus of attention for decades in personality and social psychology. Most recently, Webster and Kruglanski (1994) have developed and validated a psychometric tool—the *need for closure* scale—that effectively captures the essence of this dimension of individual variation.

We employed this scale to assess whether a press for coherence as assessed by need for closure underlies individual differences in the tendency to show proportional (linear) versus catastrophic (nonlinear) changes in behavior in response to escalation and de-escalation of provocative behavior in a social relationship. Among people with low need for closure, there is sensitivity to nuance, an appreciation of how different factors are relevant to understanding an issue, and a willingness to suspend judgment until pertinent facts and ideas are considered. In short, low NFC people maintain relatively complex representations of relationships, even those marked by conflict. Among people with a high need for closure, however, there is intolerance of nuance and little patience for having all the pertinent facts and ideas assembled and considered. In short, high NFC people are inclined to maintain relatively simple representations of relationships—perhaps especially so when a relationship is marked by conflict.

Accordingly, we hypothesized that low need for closure (NFC) participants would display a linear scenario in the context of conflict escalation, whereas their high NFC counterparts would display a nonlinear pattern. A primary consequence of this hypothesized difference is that high NFC individuals, when locked in escalated conflict, will tend to show more resistance to responding constructively to conciliatory gestures than will low NFC individuals, therefore contributing to the maintenance of hostility and conflict.

We recruited graduate students from a large university in the Northeastern United States and sent individual emails to them containing a link to an online survey. The study consisted of two sections. The first section required participants to read 11 sequential scenarios describing an interpersonal conflict situation, each involving the participant and another person in a collaborative work setting. The first 6 scenes portray progressively negative actions by the work partner, from mild and private criticism of the participant (Scene 1) to extreme and public insults of the participant (Scene 6). The subsequent scenes portray progressively conciliatory actions by the work partner, from an apology (Scene 7) to assuming the majority of the work on the collaborative task (Scene 11). After each scene, participants were provided a list of 30 possible behaviors and asked to indicate which of these they would be most likely to enact in response to the actions of the work partner. The 30 behaviors, grouped into 6 categories, represent incremental gradations on a positive vs. negative dimension. The first category (behaviors 1–5) is clearly positive (e.g., empathizing, humor), while the sixth category (behaviors 26–30) is clearly negative (e.g., taking revenge, physical aggression). The response options were randomly ordered for each scenario to avoid any systematic errors in presentation order. The 11 scenarios and 30 response options were adopted from the study by Bui-Wrzosinska (2005). The second section of the online survey assessed participants' need for closure (Webster & Kruglanski, 1994).

We hypothesized that high NFC participants would display resistance to change in their behavioral options in the face of increasing provocation by the other person, until a threshold was reached, at which time they would switch from a relatively conciliatory behavioral choice to a more aggressive choice. When the other person began to de-escalate his or her provocation and adopt an increasingly conciliatory stance, high NFC participants would persist in their aggressive response choice until a threshold was reached, at which point they would switch to less aggressive and more conciliatory behavioral choices. The two thresholds for these nonlinear changes were predicted to be different, however, in line with the hysteresis associated with nonlinear systems (and observed in Bui-Wrozinska, 2005).

The results provided support for these hypotheses. High NFC participants tended to display mild reactions to provocation at the outset and to maintain this orientation in the face of increasing provocation by the other person. However, when a threshold of provocation was reached—typically, by the second or third scenario—high NFC participants switched to higher levels of aggressive retaliation (category 6). During the de-escalation phase (scenarios 7 to 11), meanwhile, the high NFC participants tended to display a greater decrease in conflict intensity than did the low NFC participants. For the latter participants, conflict intensity tended to remain at relatively low levels (1 and 2) throughout both the escalation and de-escalation phases.

It is important to recognize that anyone, regardless of their characteristic need for closure, can develop representations of a conflict that vary in their relative complexity. Social psychological research has identified a variety of factors—stress, time pressure, and cognitive load, for example—that can promote a global, undifferentiated representation of other people, outgroups, events, and situations characterized by tension and conflict. Indeed, simply priming people to think globally about the larger meaning of an event as opposed to the details comprising the event can induce global (high-level) as opposed to differentiated (low-level) representations of the event (cf. Vallacher & Wegner, 2012). We are currently taking this approach in investigating the likelihood of observing linear versus nonlinear scenarios of conflict escalation. Pending the results obtained, this line of research may prove useful in understanding the conditions that promote dramatic spikes in conflict and in establishing contexts for conflict negotiation that can minimize such spikes.

Summing Up and Looking Ahead

This chapter breaks the law of averages. That is, it challenges the unwritten law inherent to most research on conflict in the social sciences that looks to averages in people's reported cognition, affect, and behavior for answers to the problems of conflict and peace. In its place, we have advocated for the promise of patterns. We suggest that the time is ripe to extend our understanding of social conflict beyond the moments captured by most surveys and lab studies, and to begin to investigate conflict dynamics in their natural habitat—in the context of the ups and downs, ins and outs, and wins and losses of social relations as they evolve in time. In many ways, this emphasis on temporal flow harkens back to the thinking of many of the founders of the field of social psychology such as James, Cooley, Mead and Lewin. However, with the benefits bestowed upon us by contemporary applied mathematics and computer science, we today have the tools that allow us to track and investigate conflict dynamics with much more precision and rigor.

The ability to take time seriously that comes with dynamical-systems models and methods of conflict research provides us with a basic building block necessary for understanding conflict dynamics of all kinds. Whether it is the impact over time of amazingly small differences in the initial conditions of cooperative versus competitive goals on happiness in conflict, the importance of adaptation in one's choice of strategies and tactics in unfolding disputes, the value of an optimal balance of prevention and promotion orientations in evolving struggles, or the role of a few basic parameters in determining incremental versus catastrophic escalation dynamics, one thing is certain—patterns matter. Without the capacity to observe and analyze these core aspects of conflict as they play out over time, our understanding would remain splintered, piecemeal, and very possibly mistaken.

Our task now is to turn the light of dynamical-systems models and methods of conflict research onto one of the darkest corners of social relations: protracted, intractable conflicts. The remainder of the book will take up this task.

References

Anderson, N. H. (1981). *Foundations of information integration theory*. New York, NY: Academic.

Asch, S. E. (1946). Forming impressions of personalities. *Journal of Abnormal and Social Psychology, 41*, 258–290.

Appelt, K. C., & Higgins, E. T. (2010). My way: How strategic preferences vary by negotiator role and regulatory focus. *Journal of Experimental Social Psychology, 46*, 1138–1142.

Bar-Tal, D. (2000). From intractable conflict through conflict resolution to reconciliation: Psychological analysis. *Political Psychology, 21*, 351–365.

Bui-Wrzosinska, L. (2005). The dynamics of conflict in a school setting. Unpublished Masters thesis, Warsaw School for Social Psychology.

Callanan, G. A., Benzing, C. D., & Perri, D. F. (2006). Choice of conflict-handling strategy: A matter of context. *The Journal of Psychology, 140*, 269–288.

Canary, D. J., Cupach, W. R., & Messman, S. J. (1995). Relationship conflict: Conflict in parent–child, friendship, and romantic relationships. In C. Hendrick & S. Hendrick (Series Ed.), Series on close relationships. Newbury Park, CA: Sage

Capra, F. (2002). *The hidden connections: A science for sustainable living*. London: Flamingo.

Coleman, P. T. (2003). Characteristics of protracted, intractable conflict: Toward the development of a metaframework-I. *Peace and Conflict: Journal of Peace Psychology, 9*, 1–37.

Coleman, P. T., Bui-Wrzosinska, L., Vallacher, R., & Nowak, A. (2006). Approaching protracted conflicts as dynamical systems: Guidelines and methods for intervention. In A. Schneider & C. Honeyman (Eds.), *The negotiator's fieldbook* (pp. 61–74). Chicago, IL: American Bar Association Books.

Coleman, P. T., Fisher-Yoshida, B., Stover, M., Hacking, A., & Bartoli, A. (2008). Reconstructing ripeness II: Models and methods for fostering constructive stakeholder engagement across protracted divides. *Conflict Resolution Quarterly, 26*, 43–69.

Coleman, P. T., Mitchinson, A., & Kugler, K. G. (2009). Adaptation, integration, and learning: The three legs of the steady stool of conflict resolution in asymmetrical power relations. Poster presented at the 22nd Conference of the International Association for Conflict Management, Kyoto, Japan.

Coleman, P. T., Kugler, K. G., Mitchinson, A., Chung, C., & Musallam, N. (2010). The view from above and below: The effects of power and interdependence asymmetries on conflict dynamics and outcomes in organizations. *Negotiation and Conflict Management Research, 3*, 283–311.

Coleman, P. T., & Kugler, K. G. (2011). Tracking conflict adaptivity: Introducing a dynamic measure of adaptive conflict management in organizations (WOP Working Paper No. 2011/4). Retrieved from http://www.psy.lmu.de/wirtschaftspsychologie/forschung/working_papers/index.html

Coleman, P. T., Kugler, K. G., Bui-Wzosinska, L., Nowak, A., & Vallacher, R. (2012). Getting down to the basics: A situated model of conflict in social relations. *Journal of Negotiation, 28*, 7–43.

Coleman, P. T., Kugler, K. G., & Vallacher, R. R. (2013). Hoping for the Best, Preparing for the Worst: Regulatory Focus Optimality in Social Conflict. *WOP Working Papers No. 2013 / 3*. Retrieved from: www.psy.lmu.de/wirtschaftspsychologie/forschung/working_papers/index.html

Coleman, P. T., Vallacher, R. R., Bartoli, A., Nowak, A., & Bui-Wrzosinska, L. (2011). Navigating the landscape of conflict: Applications of dynamical systems theory to addressing protracted conflict. In D. Korppen, N. Ropers, & H. J. Giessmann (Eds.), *The non-linearity of peace processes: Theory and practice of systemic conflict transformation* (pp. 39–56). Opladen, Germany: Barbara Budrich.

Coleman, P. T., Vallacher, R., Nowak, A., & Bui-Wrzosinska, L. (2007). Intractable conflict as an attractor: Presenting a dynamical model of conflict, escalation, and intractability. *The American Behavioral Scientist, 50*, 1454–1475.

Davis-Stober, C. P., Dana, J., & Budescu, D. V. (2010). Why recognition is rational: Optimality results on single-variable decision rules. *Judgment and Decision Making, 5*, 216–229.
De Cremer, D., Brebels, L., & Sedikides, C. (2008). Being uncertain about what? Procedural fairness effects as a function of general uncertainty and belongingness uncertainty. *Journal of Experimental Social Psychology, 44*, 1520–1525.
De Dreu, C. K. W. (1997). Productive conflict: The importance of conflict management and conflict issue. In C. K. W. De Dreu & E. Van de Vliert (Eds.), *Using conflict in organizations* (pp. 9–22). London: Sage.
De Dreu, C. K. W. (2010). Social conflict: The emergence and consequences of struggle and negotiation. In S. T. Fiske, D. T. Gilbert, & G. Lindzey (Eds.), *Handbook of social psychology* (5th ed., pp. 983–1023). San Francisco, CA: Wiley.
Deutsch, M. (1973). *The resolution of conflict: Constructive and destructive processes*. New Haven, CT: Yale University Press.
Deutsch, M. (1977). *The resolution of conflict: Constructive and destructive processes*. New Haven: Yale University Press.
Deutsch, M. (1982). Interdependence and psychological orientation. In V. Derlega & J. L. Grzelak (Eds.), *Cooperation and helping behavior: Theories and research* (pp. 15–42). Orlando, FL: Academic Press.
Deutsch, M. (1985). *Distributive justice: A social psychological perspective*. New Haven, CT: Yale University Press.
Deutsch, M. (2002). Social psychology's contributions to the study of conflict resolution. *Negotiation Journal, 18*, 307–320.
Deutsch, M. (2006). Cooperation and competition. In M. Deutsch, P. T. Coleman, & E. C. Marcus (Eds.), *The handbook of conflict resolution: Theory and practice* (pp. 23–42). San Francisco, CA: Wiley.
Deutsch, M. (2007). Two important but neglected ideas for social psychology as they relate to social justice. Paper presented at the Conference on Social Justice, New York.
Deutsch, M. (in press). A theory of cooperation and competition and beyond. In Kruglanski, A. E. & Higgins, E. T. (Eds.), *Theory in social psychology*.
Deutsch, M., Coleman, P. T., & Marcus, E. C. (Eds.). (2006). *Handbook of conflict resolution: Theory and practice* (2nd ed.). San Francisco, CA: Jossey-Bass.
Euwema, M. C., Van de Vliert, E., & Bakker, A. B. (2003). Substantive and relational effectiveness of organizational conflict behavior. *International Journal of Conflict Management, 14*, 119–139.
Fisher, R. J., & Keashly, L. (1990). A contingency approach to third party intervention. In R. J. Fisher (Ed.), *The social psychology of intergroup and international conflict resolution* (pp. 234–238). New York: Springer-Verlag.
Fiske, S. T., & Taylor, S. E. (1991). *Social cognition* (2nd ed.). New York: McGraw-Hill.
Galinsky, A. D., Leonardelli, G. J., Okhuysen, G. A., & Mussweiler, T. (2005). Regulatory focus at the bargaining table: Promoting distributive and integrative success. *Personality and Social Psychology Bulletin, 31*, 1087–1098.
Gersick, C. J. G. (1991). Revolutionary change theories: A multilevel exploration of the punctuated equilibrium paradigm. *The Academy of Management Review, 16*, 10–36.
Goertz, G., & Diehl, P. F. (1992). The empirical importance of enduring rivalries. *International Interactions, 18*, 151–163.
Goertz, G., & Diehl, P. F. (1993). Enduring rivalries: Theoretical constructs and empirical patterns. *International Studies Quarterly, 37*, 147–171.
Gottman, J. M., Murray, J. D., Swanson, C. C., Tyson, R., & Swanson, K. R. (2002). *The mathematics of marriage*. Cambridge, England: MIT Press.
Higgins, E. T. (1997). Beyond pleasure and pain. *American Psychologist, 52*, 1280–1300.
Higgins, E. T. (2000). Making a good decision: Value from fit. *American Psychologist, 55*, 1217–1230.

Higgins, E. T. (2012). *Beyond pleasure and pain: How motivation works*. New York, NY: Oxford University Press.

Holland, J. (1995a). Can there be a unified theory of complex adaptive systems? In H. J. Morowitz & J. L. Singer (Eds.), *The mind, the brain, and complex adaptive systems*. Boston: Addison-Wesley.

Holland, J. H. (1995b). *Emergence: From chaos to order*. Reading, MA: Addison-Wesley.

Jehn, K. A. (1995). A multimethod examination of the benefits and detriments of intragroup conflict. *Administrative Science Quarterly, 40*, 256–282.

Jehn, K. A. (1997). A qualitative analysis of conflict types and dimensions in organizational groups. *Administrative Science Quarterly, 42*, 530–557.

Johnson, D. W., Johnson, R. T., & Smith, K. A. (2000). Constructive controversy: The educative power of intellectual conflict. *Change: The Magazine of Higher Learning, 32*(1), 28–37.

Jones, W., & Hughes, S. H. (2003). Complexity, conflict resolution and how the mind works. *Conflict Resolution Quarterly, 20*, 4–20.

Kriesberg, L. (2003). *Constructive conflicts: From escalation to resolution* (2nd ed.). Lanham, MD: Rowman and Littlefield.

Kelley, H. H. (1997). The "stimulus field" for interpersonal phenomena: The source of language and thought about interpersonal events. *Personality and Social Psychology Review, 1*, 140–169.

Kriesberg, L. (2005). Nature, dynamics, and phases of intractability. In C. A. Crocker, F. O. Hampson, & P. Aall (Eds.), *Grasping the nettle: Analyzing cases of intractable conflict* (pp. 65–98). Washington, DC: United States Institute of Peace.

Kristjánsson, K. (2009). Putting emotion into the self: A response to the 2008 journal of moral education special issue on moral functioning. *Journal of Moral Education, 38*, 255–270.

Kugler, K. G., & Coleman, P. T. (2010). *Independence, conflict and autonomy: An experimental study*. Unpublished Manuscript

Latané, B., & Nowak, A. (1994). Attitudes as catastrophes: From dimensions to categories with increasing involvement. In R. R. Vallacher & A. Nowak (Eds.), *Dynamical systems in social psychology* (pp. 219–249). San Diego: Academic Press.

Liebovitch, L. S., Naudot, V., Vallacher, R., Nowak, A., Bui-Wrzosinska, L., & Coleman, P. T. (2008). Dynamics of two-actor cooperation–competition conflict models. *Physica A, 387*, 6360–6378.

Liebovitch, L. S., Vallacher, R., & Michaels, J. (2010). Dynamics of cooperation-competition interaction model. *Peace and Conflict: The Journal of Peace Psychology, 16*(2), 175–188.

Losada, M., & Heaphy, E. (2004). The role of positivity and connectivity in the performance of business teams: A nonlinear dynamics model. *The American Behavioral Scientist, 47*, 740–765.

Magee, J. C., & Galinsky, A. D. (2008). Social hierarchy: The self-reinforcing nature of power and status. *Academy of Management Annals, 2*, 351–398.

Mandela, N. (1994). *Long walk to freedom: The autobiography of Nelson Mandela*. Boston, MA: Little Brown & Co.

McClelland, D. C. (1975). *Power: The inner experience*. New York, NY: Irvington.

Mischel, W. (1977). The interaction of person and situation. In D. Magnusson & N. S. Endler (Eds.), *Personality at the crossroads: Current issues in interactional psychology* (pp. 333–352). Hillsdale, NJ: Lawrence Erlbaum Associates.

Michaels, J. L. (2012). PhD thesis, Department of Psychology, Florida Atlantic University, Boca Raton, FL.

Nelson, J. D. (2009). Naïve optimality: Subjects' heuristics can be better motivated than experimenters' optimal models. *The Behavioral and Brain Sciences, 32*, 94–95.

Nowak, A., Vallacher, R. R., Tesser, A., & Borkowski, W. (2000). Society of self: The emergence of collective properties in self-structure. *Psychological Review, 107*, 39–61.

Pollock, J. L. (2006). Against optimality: Logical foundations for decision-theoretic planning in autonomous agents. *Computational Intelligence, 22*, 1–25.

Pruitt, D. G., Kim, S. H., & Rubin, J. Z. (2004). *Social conflict: Escalation, stalemate, and settlement* (3rd ed.). Boston: McGraw-Hill.

Pruitt, D. G. (2006). A graphical interpretation of escalation and de-escalation. Presented at dynamics and complexity of intractable conflicts, Kamimierz, Poland, October 19–22, 2006

Pruitt, D. G., & Kim, S. H. (2004). *Social conflict: Escalation, stalemate, and settlement*. New York, NY: McGraw-Hill.

Reilly, B. (2001). *Democracy in divided societies*. Cambridge, England: Cambridge University Press.

Reykowski, J. (2006). Deliberative democracy and human nature: An empirical approach. *Political Psychology, 27*, 323–346.

Rosenberg, S., & Sedlak, A. (1972). Structural representations of implicit personality theory. In L. Berkowitz (Ed.), *Advances in experimental social psychology* (Vol. 6, pp. 235–297). San Diego, CA: Academic Press.

Tamura, R., Kameda, T., & Fukano, H. (2006). The robustness of Pareto-optimality in group decision-making: Do people pursue distributive justice over social efficiency? (II). *The Japanese Journal of Social Psychology, 21*, 233–240.

Ten Velden, F. S., Beersma, B., & de Dreu, C. K. W. (2009). Goal expectations meet regulatory focus: How appetitive and aversive competition influence negotiation. *Social Cognition, 27*, 437–454.

Thom, R. (1975). *Structural stability and morphogenesis*. New York, NY: Addison-Wesley.

Thorndike, E. L. (1920). A constant error in psychological ratings. *The Journal of Applied Psychology, 4*, 25–29.

Vallacher, R. R., & Nowak, A. (Eds.). (1994a). *Dynamical systems in social psychology*. San Diego, CA: Academic.

Vallacher, R. R., & Nowak, A. (1994b). The stream of social judgment. In R. R. Vallacher & A. Nowak (Eds.), *Dynamical systems in social psychology* (pp. 251–275). San Diego, CA: Academic.

Vallacher, R. R., Nowak, A., & Kaufman, J. (1994). Intrinsic dynamics of social judgment. *Journal of Personality and Social Psychology, 67*, 20–34.

Vallacher, R. R., & Wegner, D. M. (2012). Action identification theory. In P. A. M. Van Lange & A. W. Kruglanski (Eds.), *Handbook of theories of social psychology* (pp. 327–348). Thousand Oaks, CA: Sage.

Van Lange, P. A. M., Agnew, C. R., Harinck, F., & Steemers, G. E. M. (1997). From game theory to real life: How social value orientation affects willingness to sacrifice in ongoing close relationships. *Journal of Personality and Social Psychology, 73*, 1330–1344.

Van de Vliert, E., Euwema, M. C., & Huismans, S. E. (1995). Managing conflict with a subordinate or a superior: Effectiveness of conglomerated behavior. *Journal of Applied Psychology, 80*, 271–281.

Webster, D. M., & Kruglanski, A. W. (1994). Individual differences in need for cognitive closure. *Journal of Personality and Social Psychology, 67*, 1049–1062.

Wegner, D. M., & Vallacher, R. R. (1977). *Implicit psychology: An introduction to social cognition*. New York, NY: Oxford University Press.

Wish, M., Deutsch, M., & Kaplan, S. J. (1976). Perceived dimensions of interpersonal relations. *Journal of Personality and Social Psychology, 33*, 409–420.

Zartman, I. W., & Rubin, J. Z. (Eds.). (2002). *Power and negotiation*. Ann Arbor, MI: University of Michigan Press.

Chapter 5
Traps: Intractable Conflict as a Dynamical System

Although uncommon, when the perfect storm of intractable conflict hits, it can undermine the security and well-being of families, communities, and societies everywhere. When viewed through the lens of dynamical systems, both the low frequency and highly destructive consequences of this perplexing phenomenon become understandable. In general terms, conflict intractability develops when social, psychological, and structural elements interact over time to promote the emergence of a stable and coherent pattern of thought and behavior organized around perceived incompatibilities. These patterns function as attractors in that they "attract" the mental and behavioral dynamics of the parties to the conflict. As the conflict becomes a primary focus of each party's thoughts, feelings, and actions, a wide variety of events and forces—even those that seem irrelevant to the conflict—are framed in a way that intensifies or maintains the conflict. Metaphorically, the conflict acts like a gravity well into which the surrounding mental, behavioral, and social-structural landscape begins to slide. Once trapped in such a well, escape requires tremendous energy and commitment and thus becomes difficult, if not seemingly impossible. Such dynamics are today apparent in Israel/Palestine, the Colombia/FARC conflict, and even in the current low-intensity conflict dynamics on the island of Cyprus.

Our aim in this chapter is to unpack this general idea in terms of the principles and concepts introduced in the preceding chapters. We begin by describing social relations in terms of attractor dynamics, and using this perspective to illustrate how intractable conflict represents a special form of social dynamics in which basic processes line up in an unlikely but potentially destructive manner. Not all conflicts become intractable, so we discuss the dynamical processes that distinguish this potentially destructive form of social relations. The focus in particular is on the shifting balance between reinforcing and inhibiting feedback loops (relationships between variables that amplify their original states or inhibit it) in the progression from benign to malignant social relations.

With this perspective in hand, we describe three lines of research designed to assess its validity and relevance to conflict in real-world contexts. First, we present empirical studies designed to illustrate the emergence and expression of conflict

R.R. Vallacher et al., *Attracted to Conflict: Dynamic Foundations of Destructive Social Relations*, Peace Psychology Book Series, DOI 10.1007/978-3-642-35280-5_5,

intractability in interpersonal contexts when moral values provide the focus of interaction. We then discuss how the dynamics of intractable conflict can transform the social relationships in an entire society, and present preliminary evidence from field research in support of this scenario of spreading pervasiveness. The third line of research also focuses on the spread of conflict in social systems, but it is based on a mathematical model that is implemented in computer simulations. This model, which exemplifies the approach of dynamical minimalism, is based on a few assumptions introduced by Morton Deutsch (1973) in his Crude Law of Social Relations.

The final section discusses the distinction between open and closed systems, and relates this distinction to the dynamics associated with intractable conflict. We argue that whether a social system characterized by the potential for conflict can escape the pull of intractability depends on the degree to which the system is impervious to outside influences and new circumstances (closed) or instead is capable of incorporating and adapting to influences and changing circumstances (open). We conclude the chapter on an optimistic note by suggesting that the secret to transforming intractable conflicts can be found within the very framework—dynamical systems theory—that gives rise to intractability.

Intractable Conflict as a Dynamical System

Most social conflicts are temporary and tend to be resolved without devolving into violence and destruction. Indeed, as emphasized from the outset, conflict is largely a constructive process that promotes compromise and innovation at all levels of social reality, from friendship formation and marriage to representative democracy and international relations. Still, a small proportion of conflicts—about 5–7 %—tend to become protracted and appear impervious to resolution (Coleman, 2011). Like constructive conflicts, conflicts characterized by intractability can be observed at all levels of social reality—from dysfunctional marriages to political polarization and interstate warfare.

An intractable conflict seems paradoxical with respect to its dynamics. On the one hand, it is characterized by considerable dynamism, with important events unfolding on various timescales. On the other hand, an intractable conflict is characterized by long-term stability, with the parties to the relationship seemingly frozen in their mutual feelings and action orientations. The simultaneous expression of forces promoting intense dynamism and equally intense resistance to change suggests that intractable conflict can be investigated through the lens of nonlinear dynamical systems. In particular, the defining features of conflict can be recast in terms of key concepts such as self-organization, emergence, hysteresis, nonlinear transitions, and attractor dynamics.

Attractor Dynamics in Intractable Conflict

Once initiated, intractable conflicts seem to have a life of their own. Even if the specific issues that generated the conflict are resolved, the animosity between the parties to the conflict may persist or even intensify over time. A host of new issues may arise and fade in salience, but the negative perceptions, beliefs, feelings, and actions defining the relationship between the parties show little if any change. It is hard to understand such intransience in terms of the usual motivational underpinnings of social life. A conflict with no end in sight hardly serves the interests of the parties to the conflicts: to the contrary, it drains their respective resources, wastes energy, and diminishes human capital. Faced with such futility and dysfunction, a compromise that partially addresses the needs and interests of the parties should be embraced once they realize that such a compromise represents a far better deal than does maintaining a self-defeating pattern of behavior with no end in sight.

The persistence of persistent negative mental and behavioral patterns can be couched in terms of attractor dynamics. An attractor, as noted in preceding chapters, is a restricted range of states toward which a dynamical system evolves over time, and to which the system returns after it has been perturbed. Attempts to move the system out of its attractor promotes forces that reinstate the system at its attractor. Viewed through this lens, intractable conflict represents the emergence of strong, stable attractors reflecting stable patterns of thought, feeling, and action on the part of group members. Such an attractor, in effect, "attracts" a wide variety of mental and behavioral states to a narrow set of negative and destructive states.

The Attractor Landscape

The attractor concept and the relevance of attractors for conflict can be represented in a simple metaphor, depicted in Fig. 5.1. The valley represents the system's attractor and the ball represents the current state of the system. From its current position, the ball will roll down the hill and come to rest at the bottom of the valley.

The system portrayed in Fig. 5.1 has two attractors (A and B), each with its own basin of attraction, which is represented by the width of each valley. Because the basin of attraction for Attractor A is wider than the basin of attraction for Attractor B, a wider range of states will evolve toward Attractor A than toward Attractor B. The two attractors also vary in their strength, which is depicted as the respective depth of each valley. Note that B is stronger than A—that is, it has a deeper valley. In terms of the gravity metaphor, a system captured by attractror B is more difficult for it to be dislodged by external influence than is a system captured by attractor A.

These basis properties of attractors have clear relevance for the intractability of conflict. The width of the basin of attraction represents the range of ideas and actions that are likely to converge on the predominant mental and behavioral pattern characterizing the system. For a negative attractor with a wide basin of

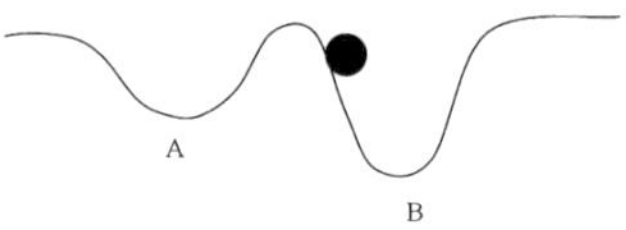

Fig. 5.1 A dynamical system with two attractors (A and B)

attraction, even positive information that contradicts the negative view of another person or group is likely to be transformed by a variety of cognitive mechanisms in line with the predominant view (cf. Nowak, Vallacher, Tesser, & Borkowski, 2000). Thus, a peaceful overture or gesture by an opposing party may become reframed so that it provides evidence in support of, rather than in contradiction to, the predominant negative response tendency of the other party.

The depth of an attractor, meanwhile, represents how difficult it is for the system to escape the pull of malignant thoughts and behaviors. When a system is at a deep attractor for negative relations (as in Attractor B), any attempt to resolve the conflict corresponds to pushing the ball uphill. This not only requires considerable effort, but it is also likely to be futile—once the pushing force is relaxed, the ball will roll back to the attractor. This scenario is illustrated by the experience of Poland during the transition from communism to democracy in the late 1980s. The Polish police attempted to suppress people's expression of political opinions, economic needs, and general dissatisfaction, but their efforts proved to be ineffective. Each police action, in fact, only served to fuel opposition and the will to resist. In effect, these actions were tantamount to pushing the ball up the hill. Because the attractor for anti-government sentiment was very strong, the police actions proved to be counter-productive, as they tended to activate forces contrary to the police efforts.

The gravity metaphor, however, suggests that efforts to move a system out of its attractor can prove successful if sufficient force is applied and if another attractor is available. Thus, the system depicted in Fig. 5.1 can escape Attractor B and settle in Attractor A if enough force is employed to move the ball up the hill and into the valley representing Attractor A's basin of attraction. This simple has important implications for change in a system characterized by intractable conflict. In particular, if a system has more than one attractor, the mental, affective, and behavioral states associated with the system tend to sort themselves categorically, with each category corresponding to a different attractor. If change occurs it such a system, it is likely to occur in an abrupt, nonlinear, qualitative manner rather than in a slow, linear, and incremental fashion (cf. Latané & Nowak, 1994).

This nonlinear scenario of change may be asymmetric, however, with many mechanisms aligned to maintain a protracted conflict. Once the conflict reaches a certain threshold, then, the attractor of destructive conflict is likely to be considerably stronger than any other attractor for the system. A considerably stronger force may therefore be required to escape the conflict than was required to drag the parties into the conflict. Moreover, even if a conflict appears to be resolved, the strong attractor of conflict continues to represent a latent danger that can recapture the mental, affective, and behavioral dynamics of the parties involved, creating the potential for a rapid reinstatement of the conflict. The rapid regression to a negative attractor was evident, for example, in the resumption of hostilities in the Middle East after the collapse of the Oslo accords in the 1990s.

Manifest and Latent Attractors

Because a system spends most of its time in the vicinity of its current attractor, the nature of this attractor can be inferred from observation of the system's pattern of behavior. The system's behavior, however, may not reveal the existence of other attractors for the system's dynamics that are currently invisible. These *latent attractors*, which specify an alternative range of possible behaviors for the system, become visible and define the intrinsic tendencies of the system when the system falls within their respective basins of attraction. Hence, the creation and destruction of latent attractors can prove critical in shaping how the system will respond to conditions and events that have yet to take place.

The existence of a latent attractor can set the stage for dramatic and unexpected changes in social relations mired in conflict. Genocides provide an extreme example of the power of latent attractors. They are typically initiated in a very rapid fashion, but they are invariably preceded by a period of changes in inter-group attitudes and behaviors, often involving the humiliation of group members, negative stereotyping, and a denial of rights. These changes are not necessarily dramatic by themselves, so opposition to them may be fairly minimal and contained. Nonetheless, each change may create and then deepen a latent attractor that can subsequently determine the fate of inter-group relations. Once an attractor for conflict has been created, even relatively small provocations can move the system into the basin of this attractor, with the entire system dragged into full-blown destructive conflict. The assassination of Archduke Ferdinand in Sarajevo in 1914, for example, probably would not have provoked World War I if a latent attractor for intense conflict had not been created by a configuration of international interests and a series of prior events.

One can also envision the converse scenario. Thus, a series of positive acts by one group toward another group may be rejected at face value and have little or no immediate effect on the relations between the groups. But such acts may set the stage for a sudden escape from a negative attractor by creating a positive latent attractor to which the system can move. Consider, for example, the Orange revolution in Ukraine, during which the relations between Poland and Ukraine changed over the course of a few weeks from fairly negative to very positive. Although the observable change was rapid, it was made possible by the positive interactions and exchanges between various groups of people that took place many years before the Orange revolution. Another example is provided by the rapid repair in relations between the U.S. and the USSR after decades of the Cold War. This transition from negative to fairly positive relations was likely fostered by a series of low-profile citizen exchanges, such as the Dartmouth conferences that occurred in the years leading up to Perestroika (Saunders, 1997).

The potential for latent attractors has important implications for the fate of social relations characterized by long-term and seemingly intractable conflict (e.g., Coleman, Bui-Wrzosinska, Vallacher, & Nowak, 2006; Coleman, Vallacher, Nowak, & Bui-Wrzosinska, 2007; Nowak, Vallacher, Bui-Wrzosinska, & Coleman, 2007; Vallacher, Coleman, Nowak, & Bui-Wrzosinska, 2010). Factors such as objectification, dehumanization, and stereotyping of outgroup members are clearly preconditions for the development of intractable conflict (Deutsch, 1973), but they not have an immediate impact on intergroup relations. However, they may gradually create a latent attractor to which the system can abruptly switch in response to a relatively minor provocation. Conversely, attempts to resolve conflict may be ineffective in the short run but nonetheless create a latent attractor for positive intergroup relations, thereby establishing a potential relationship to which the groups can switch if other conditions permit. Even between groups with a long history of conflict, a latent positive attractor can promote a rapid de-escalation of conflict.

In sum, noteworthy and relatively sudden changes in a social system correspond to changes in the attractor landscape of the system. Although a change within a basin of attraction is likely to be short-lived and resisted by the system, a change in the salience of a system's attractors can reconfigure the ensemble of likely states that can be adopted and thus alters the forces operating within the system. Changes in a system's attractor landscape can promote changes in behavior that are become self-sustaining and self-enhancing over time.

Levels of Social Reality and Attractor Dynamics

Because attractors represent formal properties of dynamical systems, they characterize the dynamics of systems at all levels of personal and social reality (cf. Nowak & Vallacher, 1998; Vallacher & Nowak, 2007). So although cognitive processes and inter-group relations are very different phenomena, for example, both can be described in terms of stable patterns (e.g., interpretation in cognition, social distance in inter-group relations) that provide coherence and stable platforms for action. This suggests that conflict can be characterized with attractors at different levels of social reality. Attractors at the level of the mind represent stable patterns of thought and affect concerning the parties to the conflict (e.g., oneself and another person in interpersonal conflict, the ingroup and outgroup in inter-group conflict). Attractors at the interpersonal and inter-group levels, meanwhile, represent stable patterns of relations between the parties to the conflict.

There is feedback among levels in a dynamical system, so an attractor that initially develops at one level of reality is likely to forge attractors at other levels. Two groups may come into conflict over issues of resource allocation, for example, and the resultant negative interactions could promote psychological attractors (i.e., negative attitudes and feelings) that are consistent with, and serve to maintain, the interpersonal conflict. Indeed, once a psychological attractor develops, the factors that spawned the conflict (i.e., resource allocation) could be subsequently resolved mitigated without resolving the conflict if the attractor is strong and has a wide basin of attraction. Thus, a change in objective conditions could be interpreted in a manner that reinforces rather than undermines the negative feelings toward the other party. In conflict scenarios, then, an interpersonal or inter-group attractor is likely to promote a social judgment attractor, which then can reinforces the interpersonal attractor, and so on, in a reciprocal reinforcing feedback loop.

Because of the potential for feedback among levels, a conflict initially associated with one level of reality can create the potential for overlapping and mutually reinforcing attractors that serve to anchor the conflict. Moreover, the expansion of attractors due to feedback among levels can become encoded in cultural beliefs and traditions, which adds yet another level to the conflict, thereby promoting long-term intransigence and resistance to attempts at conflict resolution. An example is provided by the annual *Orange walk* in Northern Ireland each July, which celebrates the Protestant victory of Prince William of Orange over King James II in 1690. This event reinforces memories and narratives of past grievances and losses associated with the protracted conflict over Ireland and thus holds potential for triggering latent psycho-social attractors for more hostile Catholic-Protestant relations.

The Emergence of Intractable Conflict

Once antagonistic relations between individuals, groups, or nations have become intractable, it is easy to appreciate how they are maintained with respect to attractor dynamics. But how do social relations get to this point in the first place? Conflict is ubiquitous and largely constructive in human experience, after all, and only a small proportion of conflicts become destructive and protracted over a long period of time. What dynamic processes transform a normally benign feature of social relations into something that is irrational and dysfunctional?

Dynamics of Self-Regulation

The development of intractable hostility in social relations can be looked upon as a breakdown in the self-regulatory mechanisms that underlie normal behavior. The capacity for monitoring, controlling, and changing one's thoughts, feelings, and actions is widely recognized as a defining feature of human functioning (cf. Baumeister & Heatherton, 1996; Carver & Scheier, 1999; Miller, Galanter, & Pribram, 1960; Powers, 1973; Vallacher & Nowak, 1999; Vallacher & Wegner, 1987; Wegner & Pennebaker, 1993; Weiner, 1948). In the dynamical framework, the self-regulatory tendencies of a system reflect the nature of the feedback loops by which elements influence one another.

Elements linked by *reinforcing feedback loops* bolster and amplify one another's current state. A favorable thought about a person or group, for example, elicits other favorable thoughts or transforms neutral thoughts into favorable ones, and these thoughts in turn reinforce the state of the original thought. Through repeated iterations of this feedback loop, the separate thoughts coalesce into a global judgment with positive valence. By the same token, reinforcing feedback can promote the emergence of a derogatory judgment if an unfavorable thought about a person or group recruits other unfavorable thoughts and influences neutral thoughts to take on the same valence. In contrast, elements linked by *inhibiting feedback loops* inhibit or reverse one another's current state. Thus, a favorable thought about a person or group recruits unfavorable thoughts or transforms neutral thoughts into unfavorable ones, whereas an unfavorable thought recruits favorable thoughts or transforms neutral thoughts into favorable ones.

Self-regulation involves both types of feedback (Vallacher & Nowak, 1999). Social judgment generated solely by reinforcing feedback loops can transform a single positive or negative thought about a person or group into a global evaluation that is highly polarized in a positive or negative direction, respectively (e.g., Tesser, 1978; Vallacher, Nowak, & Kaufman, 1994). On the other hand, if social judgment were generated solely by inhibitory feedback loops, an unequivocal evaluation of a person or group would never be attained, as every valenced thought would be contradicted by a thought with the opposite valence. Effective self-regulation thus involves a balance between reinforcing and inhibitory feedback loops.

This balance shapes the trajectory of thought and behavior in everyday social relations—including relations characterized by benign or constructive conflict. A hostile thought about a person or group, for example, elicits other negative thoughts or transforms neutral thoughts into negative ones, and through repeated iterations of this linkage, the separate thoughts coalesce into a coherent negative judgment. In most social contexts, however, this potential for the spread of negativity and conflict escalation is restrained by inhibitory feedback among elements. A hostile action toward a person, for example, might promote feelings of regret or guilt, or perhaps a consideration of the person's benign or positive qualities. Such thoughts effectively restrain the negative judgment and hostile impulses, establishing a tempered stance in the social relationship.

Achieving an optimal balance between reinforcing and inhibiting feedback loops is not an automatic consequence of mind and action, but rather requires sufficient *controlled processes*—conscious, deliberative, and reflective modes of thought—to enable self-awareness, impulse control, reconsideration of judgments and conclusions, and action monitoring (e.g., Carver & Scheier, 1999; Gilbert, 1993; Vallacher & Nowak, 1999; Wegner, 1994). Nonetheless, this mode of thought and action provides a reliable foundation for personal, interpersonal, and collective behavior in most circumstances. Even in situations of potential or manifest social conflict, the dynamics of mind and action generally reflect effective self-regulatory mechanisms. People can change their mind or reverse a course of action, provided they function in accordance with controlled processes that strike a balance between reinforcing and inhibitory feedback.

Breakdown in Self-Regulation

If balanced feedback is the norm for everyday thought and behavior, what is responsible for the breakdown in self-regulatory dynamics that allows strong attractors of negative thought, emotion, and action to develop? Self-regulatory failure is actually a fairly hot topic in contemporary social psychology. Research in several different paradigms has identified a host of factors that can undermine or deplete the executive resources necessary for controlled processes (e.g., Baumeister, Vohs, & Tice, 2007; Vallacher & Nowak, 1999; Vohs & Heatherton, 2000; Wegner, 1994) and thus tip the balance in favor of reinforcing feedback at the expense of inhibitory feedback. These factors, by extension, promote polarization in mental, affective, and behavioral states by undermining people's ability to inhibit thoughts, feelings, and actions that are biased in a particular direction.

Stress is a primary factor in self-regulatory failure because it can seriously compromise controlled processes (cf. Glass & Singer, 1972). Several lines of resea6rch demonstrate that people find it hard to think clearly or correct a course of action when they are under pressure, faced with tasks that exceed their self-perceived competence, or must decide between different courses of action with unclear consequences (e.g., Tice, Bratslavsky, & Baumeister, 2001; Vohs, Baumeister, & Ciarocco, 2005). Stress also increases people's felt *need for closure*, a desire to "seize" a global judgment without considering other perspectives and to "freeze" on that judgment and demonstrate strong resistance to disconfirming evidence and contrary ideas (Kruglanski & Webster, 1996). When people feel a heightened need for closure, they are not only highly defensive when exposed to contradictory judgments, they also demonstrate intolerance and antagonism toward people who express such judgments. Stress, of course, is inherent in prolonged and difficult conflict, so quite plausibly the pressures and tensions experienced in such conflicts contribute to a breakdown in normal self-regulation and promote the emergence of an attractor defined in terms of coherent and polarized thoughts, feelings, and actions.

A special form of stress is associated with people's awareness of their mortality, an existential state that is likely to have heightened salience in a situation with potential for violent conflict. According to *terror management theory* (TMT), people embrace the values of their culture most vigorously when they are mindful of their own death (cf. Greenberg et al., 1990). Because destructive conflicts can devolve toward violent confrontations, the death of oneself or of one's fellow group members may be considered a very real possibility. Interestingly, research on TMT has established that when people are made mindful of death, they become more hostile to people who criticize their country (Greenberg et al., 1990), more hostile to outgroups (Dechesne, Greenberg, Arndt, & Schimel, 2000), more punitive toward those who challenge established laws and procedures (Rosenblatt, Greenberg, Solomon, Pyszczynski, & Lyon, 1989), and more reluctant to use cultural artifacts such as a country's flag for a mundane purpose such as a table covering (Greenberg, Porteus, Simon, Pyszczynski, & Solomon, 1995). These findings suggest that the emergence and maintenance of strong attractors for hostility in interpersonal and intergroup relations might be a consequence of the recognition that one's life is at risk under conditions of intense conflict. The possibility of death associated with destructive violence, in other words, can tip the balance in self-regulation toward reinforcing feedback, which promotes global judgments with unequivocal valence.

Stress is not the only source of self-regulatory failure. Simply attending to multiple sources of information or engaging in parallel activities, for example, can undermine the controlled processes necessary for effective self-regulation. Under such conditions of *cognitive load*, people fail to correct unwarranted evaluative judgments of others, presumably because they do not take into account factors in the situation that may have promoted the behavior in question (Gilbert, 1989). Thus, negative actions are attributed to the actor's personality or values despite pressures or expectancies that would likely elicit the same behavior in anyone (e.g., Gilbert, Pelham, & Krull, 1988). Stereotypical beliefs, moreover, tend to predominate when judging members of outgroups if the perceiver is under cognitive load, since stereotypes are over-learned ways of thinking that do not require controlled processes of thinking (Gilbert & Hixon, 1991). Cognitive load also undermines people's ability to suppress thoughts and feelings that are unwanted or inappropriate (cf. Wegner, 1994; Wegner, Erber, & Zanakos, 1993; Wegner, Schneider, Carter, & White, 1987). A negative characterization of a person or group, for example, may be difficult to ignore or discount, even if the characterization is unfounded and should not be considered valid or appropriate.

If substantial cognitive load is experienced in the midst of a potentially destructive conflict, one might expect an uptick in shallow processing of information regarding members of an outgroup, with the balance shifted toward reinforcing feedback at the expense of inhibitory feedback. Although this possibility has not been directly tested, it seems reasonable. When a conflict begins to occupy people's thoughts, after all, everyone must still devote mental energy to the more mundane aspects of life. Adding thoughts about conflict to the controlled processing necessary to get through daily life, then, might serve to create a level of cognitive load that undermines balanced self-regulation. If so, people consumed with conflict on top of their everyday concerns might be prone to develop polarized thoughts and action tendencies toward members of the antagonistic group. The emergence of attractor dynamics in this fashion may set the stage for intractable conflict.

Self-regulatory failure may have a biological basis. Brain functions utilize a special form of glucose, called glycogen, and this "mental fuel" is burned at a particularly rapid rate when a person is engaged in the controlled processing necessary for self-regulation (Gailliot et al., 2007). Compared to participants who are given a natural sweetener that provides glycogen to the brain, for example, those who are given an artificial sweetener that lacks glycogen-producing properties are more likely to demonstrate subsequent lapses in self-regulation. Such lapses may be manifest as selfish or even immoral action, undertaken without consideration of the act's larger meaning or long-term consequences. People who inhibit an impulse or delay gratification, for example, are subsequently inclined to act selfishly rather than altruistically, and to behave dishonestly if such behavior enhances their immediate self-interest (e.g., Vohs et al., 2008).

This perspective suggests that prolonged efforts at self-regulation can result in a boomerang effect, with inhibitory feedback playing a weakened role in subsequent thought and behavior (cf. Baumeister, Bratslavsky, Muraven, & Tice, 1998; Gailliot et al., 2007). For example, in tense social relations characterized by sustained pressures for compromise and perspective taking—higher-level mental activities that deplete glycogen—there may be a sudden breakdown in inhibitory feedback, with the emergence of highly polarized thoughts and feelings shaping how the respective parties respond to one another. This is clearly a speculative scenario, but one that seems worthy of exploration. If confirmed, it would suggest an ironic prediction: heightened concern with compromise and problem solving in a conflict scenario holds a risk of transforming a benign conflict into one characterized by extreme judgments and non-compromising action orientations.

To this point, we have concentrated on generic factors in everyday life that can disrupt the balance between reinforcing and inhibitory feedback in thoughts, feelings, and actions. There are unique factors in social relations characterized by conflict, however, and these may hold potential for undermining self-regulation and setting parties to a conflict on the path toward intractability. Likely candidates include power symmetry versus asymmetry, the degree of confined interdependence, the level of cooperation versus competition over resources, the degree to which access to resources has a zero-sum nature, and the salience and magnitude of differences in ideology and social identity. The relation of these features of conflict to the emergence of intractability characterized by strong attractors for hostility has yet to be fully investigated. This issue, however, is front and center at several points in the following chapters.

A Hypothetical Scenario of the Emergence of Intractability

The dynamical systems framework can be made concrete by means of a hypothetical scenario in which a conflict escalates to intractability. In this scenario, the conflict might start with someone's thoughts concerning another person that center on incompatibility with that person (however, it could also start top-down, such as when a youth is socialized into a protracted conflict from birth). A structure begins to form as separate (negative) thoughts begin to support each other. This structure grows by assimilating a growing number of other psychological processes and emotions. As a result, judgments of the person become increasingly undifferentiated and unidimensional, organized around incompatibility with the person. Eventually, hostile intentions are likely to be communicated and initiate similar processes in the other person. At this moment, two systems of conflict existing on individual levels reinforce each other and eliminate the possibility for positive interaction. The thoughts, feelings, and behaviors of both people are processed within the structure of conflict. Conflict now exists at the interpersonal level.

As both parties seek support by recruiting other members of their respective groups, the conflict escalates on the social level, where it is sustained by links between the groups and by positive (reinforcing) feedback within each group. The sustainability of the conflict increases because even if the person who was the source of the conflict tries to disassemble the conflict structure, feedback from others in either group will reinstate the conflict. With the growth in intensity of the conflict, hostility and violence become increasingly likely as a result of the mutual feedback. The structure of conflict now permeates individuals, groups, and the whole society, recruiting almost all processes and events and thereby eliminating almost all chances of positive interactions between the opposing groups. At some point, the multitude of feedback loops within and between levels that sustain and support the conflict renders fruitless the attempt to disassemble even large parts of the conflict because the conflict would become reinstated in other parts of the structure.

Protracted existence of conflict at the societal level is likely to start shaping the symbol system of those involved in the conflict. Identities are built around incompatibility. The terms used for referencing the outgroup become dehumanizing in nature and preclude positive interactions. In effect, the protracted conflict becomes embedded in the culture of the society. When this happens, the conflict acquires a new means of maintenance and spreading. Anyone adopting the culture is likely to adopt the conflict embedded within it. People who never had contact with the outgroup members are nonetheless unlikely to form sustainable positive or even neutral relationships with the members of the outgroup. In this way, conflict is passed through generations.

Referring to intractable conflict as "malignant" (Coleman, 2003) is warranted because the analogy between the two terms is direct. Cancer works by penetrating the structure of the organism and enslaving essential elements of the body, which lose their original functions and begin working in service of the structure of cancer. The collapse of complexity associated with intractable conflict is visible in the same process. The multitude of psychological and social processes necessary for the maintenance of mental structures, religions, and societies become enslaved into the one-dimensional structure of conflict. Their original functions essentially vanish. As the machine of war is established and grows, it encompasses a growing number of social, economic, and political processes, focusing them all in a single issue within the conflict.

In such a structure, love, friendship, or even collaboration at work between members of the opposing parties will not be seen in its true nature, but rather as collaboration with the enemy, weakness, or treason. The richness and multidimensionality of all the processes occurring in a healthy society become entrained in the structure, leaving no opportunity for positive interactions. Such structures of intractable conflict can sustain multiple states, such as attack, revenge, temporary truce, and strategic withdrawal, but not co-existence of long-term cooperation.

The cost of conflict is not limited to suffering, loss, or deaths in the center of the conflict. Indeed, the most costly long-term consequences are not associated with the direct damage, but rather with the elimination of possibilities of positive events that are necessary to the functioning of any healthy society. Conflicts dramatically diminish the social capital of a society, an effect that inevitably results in a dramatic social and economic decline. Such consequences, however, may not be observable for a considerable period of time. Thus, the ability to diagnose the sustainable states of the society is likely to provide an early measure of the potential long-term damage to the society as well indicate the potential for a peace process. Such diagnosis would also provide a measure of the effectiveness of various intervention attempts before they actually impact the visible states of the conflict, and it would allow one to estimate the damage done by aggravating events before the damage is manifest.

From this perspective, the severity of intractable conflict can be characterized with respect to both structure and dynamics. Structurally, we can describe which elements are relevant and how these elements are inter-related. What is the nature of the feedback loops among elements? At what level does the structure of the conflict currently exist? What proportion of psychological and social processes are organized into the structure of conflict? What proportion remains conflict-free? Even in intense conflict, there may be some contexts and issues (e.g., sharing water resources) for which the parties are able to build relations or even find solutions independently of the conflict. The degree to which conflict permeates all psychological and social processes, thereby organizing all judgments and actions in terms of incompatibility, provides a measure of the conflict's intensity.

So while the severity of conflict may be related to the amount of violence between groups, the intractability of conflict may be defined in terms of the elimination of sustainable states of positive interaction. In dynamical terms, an intractable conflict lacks attractors for positive states. The progression, transformation, and de-escalation of intractable conflict can be described in terms of the dynamics of the ensemble of sustainable states for both positive and negative interactions—that is, in terms of the attractor landscape of the system. Attempts at conflict resolution are unlikely to succeed if they do not work toward the achievement of sustainable states. They may result in a temporary ceasefire, but not in long-term co-existence between the opposing parties. If no sustainable states exist, the first step at intervention should be directed to changing the ensemble of sustainable states. Only after such change has occurred can the system be effectively moved to a benign or positive state.

Research on the Dynamics of Intractable Conflict

The dynamical perspective on intractable conflict is intended to provide integrated understanding of a phenomenon that is highly diverse and seemingly idiosyncratic in its various manifestations. But for this perspective to be useful, it must be heuristic as well, generating new hypotheses that are open to verification. Accordingly, we have developed three lines of research reflecting our multi-disciplinary orientation. The first involves laboratory studies focused on moral conflicts, such as those over abortion and the right to take one's own life, which have a high propensity for escalating and becoming polarized and protracted. These studies provide empirical evidence of the constructive versus destructive effects of high-complexity versus low-complexity cognitive, emotional, and behavioral dynamics. The second approach is a conceptual model detailing the pervasive spread of negativity that is commonly observed in situations of protracted conflict. This model is illustrated with a qualitative study that examines the contagion effects in the ongoing conflict between Israelis and Palestinians. The third approach involves mathematical modeling and computer visualization of Deutsch's Crude Law of Social Relations (Deutsch, 1973), which has been identified by scholars as a primary mechanism at the core of escalation, de-escalation, and intractability (Nowak et al., 2010a).

High and Low Complexity Dynamics in Moral Conflicts

This research program tests the central proposition of our approach, that intractable conflicts (ones that are destructive, persistent, and resistant to change) evidence attractors that are characterized by low levels of complexity (e.g., fixed-point attractors). Other scholars have found evidence of attractors for difficult social dynamics. Research on marriage and divorce by Gottman (2002) has consistently found that the relational phase-space of couples whose relationships end in divorce tend to be characterized by relatively strong, coherent attractors for negative emotions and weak attractors for positive emotions. Similarly, Losada (1999) has found that the emotional and behavioral attractors of low-functioning business strategy groups are very different from high-functioning groups. High-functioning groups tend to display relatively complex and open patterns of communication that allow for the expression of thoughts, feelings, and actions that are negative as well as positive. In marked contrast, low-functioning groups display patterns of thought, feeling, and action that are both more constrained and more negative in tone and substance.

Both lines of research have begun to identify basic parameters that distinguish between attractors leading to more unfavorable and negative social outcomes versus more favorable and positive outcomes. The underlying assumption is that attractors that evidence more complexity and openness in human thinking, feeling, and acting are associated with more favorable social dynamics. The parameters characteristic of higher complexity and openness found in empirical research to date include higher ratios of positivity-to-negativity of emotions (Bales, 1950; Gottman, Markman, & Notarius, 1977), more balance between internally- and externally-focused references in groups (Losada, 1999; Losada & Heaphy, 2004), and greater balance between inquiry acts (questioning) and advocacy acts (positioning) in problem-solving groups (Losada & Heaphy, 2004).

To further investigate these dynamics, we developed a laboratory paradigm that allowed us to trace the temporal trajectories of constructive versus destructive discussions over seemingly intractable moral issues (for a description of the study, see Kugler, Coleman, & Fuchs, 2011). We hypothesized that destructive discussions over moral conflicts would be characterized by attractors which evidenced lower levels of complexity, whereas more constructive discussions would evidence more high-complexity patterns. In the context of these studies, complexity refers to the way humans respond to a potentially complex environment (cf. Satish, 1997). It was assumed that most moral conflicts, being inherently nuanced and multi-dimensional, require more complex responses to be managed constructively and effectively.

We conducted two studies. In both studies, dyads of previously unacquainted participants were brought together our lab to discuss an issue that held potential for generating an intractable conflict. In particular, the individuals comprising each dyad were matched according to their pre-measured opposing views on a topic such as abortion, euthanasia, or the death penalty, and then were asked to discuss the topic and write a joint position statement. Throughout the course of their interaction, we gathered data about their emotions, cognitions, and behaviors. We hypothesized that conflicts leading to more tractable versus more intractable outcomes would exhibit more complex cognitive, emotional, and behavioral dynamics during the discussion. The concept of complexity was supplemented by the concept of openness to the other party's concerns, opinions, and emotional experiences. Openness was presumed to be instrumental for gaining a more complex view of the conflict issue and a better understand the other party (e.g., Johnson & Johnson, 2006).

In Study 1, we compared dyads that had concluded their discussion with tractable outcomes (those showing a mutual and elaborated joint understanding of the issue) with dyads that had ended with more intractable outcomes (those showing no or an incompatible joint understanding of the issue). The comparison supported the hypothesis that those dyads ending with tractable outcomes exhibited more complex and open dynamics. Highly complex and open dynamics were operationalized by: (1) high levels of integrative complexity (i.e., differentiation and integration of perspectives and information) as well as openness to learning new information from the other party; (2) a high ratio of positive-to-negative emotions (i.e., positive emotions are thought to broaden the momentary action and thought repertoire and are therefore allowing for more complexity, e.g., Fredrickson, 1998, 2001, 2006), as well as a higher level of emotional connectedness (indicating an openness and responsiveness to the other party, e.g., Gottman & Silver, 1999; Gottman, Swanson, & Swanson, 2002); (3) a high pro-social/pro-self ratio (pro-social motivated behaviors are related to approaching conflict situations in more complex categories, Carnevale & Probst, 1998), as well as higher levels of inquiry/advocacy ratios (inquiry exhibits openness to the other parties thinking, e.g. Losada & Heaphy, 2004).

Study 2 was identical to Study 1 with one important change: We manipulated the level of integrative complexity of both members of the dyads prior to their discussion (high complexity dyads versus low complexity dyads). High versus low levels of integrative complexity was induced through two different versions of a written text on the specific moral issue under discussion (abortion, euthanasia, etc.): one was written according to the standards for high integrative complexity, the other according to the standards for low integrative complexity (cf. Baker-Brown et al., 1992). This manipulation had a significant impact on the conflict dynamics (cognitive, emotional, and behavioral complexity) as well as on the conflict outcomes (tractable vs. intractable). During the discussion, high-complexity dyads showed more complex thinking (higher levels of integrative complexity), and following the discussion dyads showed more complex emotions (higher positivity/negativity ratios), more complex behaviors (higher prosocial/proself ratios), and reached more tractable outcomes (a more elaborated joint position statement) than did the low complexity dyads.

However, we did not find the same effects for openness that were observed in the first study. We assume that the manipulation, which deliberately influenced cognitive complexity (integrative complexity), had a stronger and more direct impact on the complexity dynamics than on the openness dynamics. Even though we assume that the level of complexity and openness are connected, their relationship does not seem to be linear or direct: enhancing complexity did not automatically lead to more openness.

The two studies offer support for the proposition that constructive conflict processes over potentially intractable issues exhibit more complexity along with more openness in disputants' experiences and behaviors. Nevertheless, much work remains to be done. First, more research is needed to verify the connection between conflict intractability and attractor dynamics (c.f. Coleman et al., 2007; Vallacher et al., 2010; for another research example see Kugler & Brodbeck, 2011). How do these differences in conflict dynamics develop in time? How are they maintained once they are manifest? How might the combination of high and low-complexity conditions operate—for example, will dyads tend to move to the lowest or highest common denominator?

The Spread of Conflict into Intractability

Most conflicts have relatively clear boundaries that delineate what they are and are not about, who they concern and who they don't, and when and where it is appropriate to engage in the conflict. In situations where conflicts become more intractable, the experience of threat can become so basic and intense that the negative effects of the conflict spread and become increasingly pervasive, affecting many or even all aspects of a person's or a community's social and political life (Coleman, 2003; Musallam, Coleman, & Nowak, 2010; Rouhana & Bar-Tal, 1998; Zartman, 2005). The core existential nature of these conflicts can impact everything from policy-making, leadership, education, the arts, and scholarly inquiry down to the most mundane decisions such as whether to shop and eat in particular public places. The totality of such experiences can feel overwhelming.

Although hostilities may be most obvious at a particular level (e.g., between groups), intractable conflicts often affect and are affected by elements at multiple levels (from the psychological to the societal or geopolitical). These elements interact in complex ways, leading to increased chaos and confusion. Intractable conflicts also tend to attract more and more parties over time to the dispute. The original disputants then come to operate within a network of other relationships that may or may not be opposed. This can include formal and informal third parties, elites, observers, bystanders, NGOs, business organizations, etc (see Crocker, Hampson, & Aall, 1999; Diamond & McDonald, 1996; Staub, 2001). In Kashmir, for example, relations between the Indian, Pakistani, and Kashmiri leadership are dependent upon, among others, relations between the leaders and their general publics, hawks and doves within both groups, formal and informal representatives, extremist factions on both sides, the media and the public, business communities on all sides, regional neighbors, the Muslim and Hindu worlds, and the international community.

In the following account, the dynamical systems perspective is employed to enhance our understanding of how negativity in protracted conflict spreads to previously neutral issues, parties, and structures, and to suggest a need for assessing how this form of contagion occurs in conflicts over time (Musallam et al., 2010).

Pervasiveness

The destructiveness of long-term geopolitical conflicts is typically assessed by their *intensity* in quantitative terms, which typically includes one of three sets of criteria: (1) counting the number, types, and magnitude of bombing attacks, battles, fatalities, injuries, population displacement, etc.; (2) assessing their *temporal scope* or how long they persist, or (3) assessing the *direct consequences* of conflicts, which entails the examination of the physical and psychological welfare of those caught up in the dynamics of conflict, and the damage to the social, political, economic and educational infrastructure of the society.

However, in characterizing intractable conflicts, scholars such as Zartman (2005) have concluded that "intractability is a dynamic, self-reinforcing condition; digging an ever-deepening hole for itself and feeding itself like a vortex" (p. 55). In other words, the longer violent conflicts persist, the more they spread into the central fabric of societies, and therefore the longer they last. Zartman's observation on the nature of intractability captures the essence of how conflicts become pervasive: "They [conflicts] are not static; they grow, in both degree and nature, extending and defending themselves, so that efforts to overcome them must penetrate several layers and deal with their protective dynamics [. . ..] intractability feeds on intractability and grows with the feeding [. . .] the importance of protraction lies not in its numerical duration but in that duration's effect (2005, p. 49)."

Other scholars have specified particular processes and outcomes related to the pervasive spread of negativity in conflict. For example, Rouhana and Bar-Tal

(1998) described several aspects of public life that can become affected by conflict when it is highly central and "total". They write:

> When conflicts are total, they penetrate the societal fabric of both parties and force themselves on individuals and institutions. Leaders, publics, and institutions—such as educational and cultural systems—become involved in the conflicts. At some stages of the conflicts, intellectual life and scholarly inquiry become politicized as interest in the other society originates in the motivation to "know your enemy" and inquires become guided by security needs and considerations… [centrality] is reflected in the group members' preoccupation with the conflicts. Thoughts related to the conflict are highly accessible and are relevant to various discussions within each society (Bar-Tal, Raviv, & Freund, 1994). The centrality of such conflicts is further reflected in their saliency on the public agenda. The media and the political and intellectual elites are greatly preoccupied with the conflicts and their developments (p. 762).

Various conflict scholars have observed that destructive conflicts have a tendency to spread and thereby sustain their negative effects (Coleman, 2000, 2003; Gray, Coleman, & Putnam, 2007; Kriesberg, 2005; Rouhana & Bar-Tal, 1998; Zartman, 2005). Yet despite decades of research dedicated to the study of protracted conflict dynamics (see Azar, 1983, 1990; Burgess & Burgess, 1996; Coleman, 2003, 2004, 2006; Crocker & Hampson, 2004; Crocker, Hampson, & Aall, 2005; Deutsch, 1973; Druckman, 2001; Kelman, 1997; Kriesberg, 2005; Zartman, 1989), to date little is known about the basic mechanisms that account for the spread of conflict, or about the conditions that make it more or less likely to occur.

A Dynamical Model of Conflict Pervasiveness

For the purposes of this model, conflict pervasiveness is defined as the process through which the "spread of negativity into previously independent states and activities" transpires (Musallam et al., 2010). The spread of negativity is a process in which neutral states that are previously perceived as independent of a conflict begin to be associated with the conflict and contribute to intensifying it or spreading its effects further, instead of maintaining their original neutral states. For instance, in Israel/Palestine, healthcare delivery vehicles such as ambulance were once considered constructive and necessary components of functioning society, until they began to be used to transport weapons, at which point their presence took on a qualitatively different meaning. Conflict pervasiveness is dynamical by nature; it is an evolving set of cognitive, emotional, behavioral, and/or structural elements that grow in number, connectedness, and size over time. It is similar to the spread of a fire through a forest or an epidemic through a community. Therefore, it becomes necessary to adopt a conceptual framework for studying it that can help better understand these evolving, proliferating dynamics.

Biological and social models of processes such as epidemiology of diseases and contagion and diffusion of attitudes reveal the utility of using frameworks informed by complexity science, including complex systems theory, network models, and dynamical systems theory. For example, such frameworks have been fruitfully applied to the outbreak of epidemics (Morris & Kretzchmar, 1995), as well as to processes of increased stereotyping (McClelland, Rumelhart, & Hinton, 1986; Operario & Fiske, 2001) and aggression (Read & Miller, 2002; Read, Vanman, & Miller, 1997; Shoda, Tiernan, & Mischel, 2002).

To reiterate, dynamical systems theory concerns the study of complex systems that evolve over time. A dynamical system is defined as a set of elements that change over time as each element adjusts to the myriad influences from the other elements that make up the system. These elements can be thoughts, feelings, and actions at an individual level; people, groups and norms at a social level, or various institutions such as families, the media, religious organizations, schools, etc. at a broader level. Each element can be stimulated and perpetuated along its current path through *reinforcing feedback loops* between elements, where one element stimulates another along its current trajectory and this element, in turn, stimulates the first—thus making a loop. We see this when a negative act by an outgroup member links to negative memories and feelings from previous encounters and increase a general sense of animosity toward the outgroup and the likelihood that they will perceive future acts as negative. Elements can also obstruct or reverse one another via *inhibiting feedback loops* where one element constrains another (Coleman et al., 2006). We see this when a sense of guilt or compassion arises within us and buffers our hostile intentions or actions.

As noted earlier, feedback loops can operate both within and between levels of a system. For instance, the perception of an outgroup member as "evil" by an ingroup member can be reinforced psychologically by other thoughts and feelings that are consistent with this assessment, as well as socially by other ingroup members that hold and thus reinforce similar beliefs. At a group level, individuals can mutually influence and inform each other via reinforcing feedback in a process that creates increasing attitude homogeneity within subgroups resulting in increased conformity, or they can contradict and inhibit each other (inhibiting feedback) through processes such as criticism and dissent that eventually results in increased heterogeneity of opinion within groups.

Thus, the spread of malignancy (or positivity) over time in a conflict can be characterized as an increase in reinforcing feedback loops and a decrease in inhibiting feedback loops between various elements both within and between the psychological, social, and structural levels of a social system. Psychologically, loops become established between various negative conflict-related attitudes or thoughts which link with other negative emotions within an individual and begin to affect cognitions that were previously unrelated to the conflict. Under some conditions, these elements can be transformed, such as when neutral or positive thoughts about another person become consistently negative. Then, the activation of this thought by external stimuli or events will lead to the activation of other linked cognitions, thereby psychologically drawing the individual into the negativity of the conflict.

Such negative cognitive-emotional clusters have the capacity to spread beyond the individual psychological level and begin to operate (and create reinforcing loops) through behaviors on the interpersonal and group levels, and ultimately in changes on a structural level. In other words, for any individual, the likelihood that thought A leads to thought B and emotion C is usually guided by linkages and loops among cognitions and feelings available to that individual (Shoda et al., 2002), which can link to their behavioral responses to members of group D and be particularly likely within the context of situation E. What is important to emphasize is that the spread of conflict is not necessarily bottom up (individual to environment) or top down (environment to individual), but rather they can spread in either or both dirUections and be mutually reinforcing as long as the system lacks sufficient inhibiting feedback loops to mitigate such mutual reinforcement.

There is anecdotal evidence supporting the spread and displacement of aggression from one target group into other groups. Wessells (2007) noted that in Angola, a place torn by protracted civil wars, aggression became more and more acceptable as a means of solving community problems, and over time spread to other domains of life including friendships, martial relationships, and work relations. Similarly, Martin-Baro noted that during war, the slaughter of individuals, the disappearance of loved ones, and the constant threat to one's life all come to be accepted as increasingly normal despite their extreme natures (See Martin-Baro, as citied in Aron & Corne, 1994, Chaps. 6 and 7, pp. 108–135; Comas-Díaz, Lykes, & Alarcón, 1998). This normalization and spread of violence-related acts has been demonstrated statistically in many conflict areas. Reports such as the United Nations Population Fund (UNFPA) have consistently reported increased domestic violence against women (UNFPA, 2006), increases in organized and gang-related crimes such as kidnapping children (UNAMI, 2006), and increases in sexual abuse and trafficking in zones of protracted conflict (UNAMI, 2006; UNFPA, 2006).

In addition to the spread of aggression, conflicts are usually characterized by contagion or "spillover" of attitudes of intolerance toward outgroups or subgroups that are not part of the ingroup with respect to a particular conflict (Shamir & Sagiv-Schifter, 2006; Sullivan, Shamir, Walsh, & Roberts, 1985). For instance, in conflict zones there are often reports of increased intolerance towards other ethnic and religious minority groups (Shamir & Sagiv-Schifter, 2006), increased political intolerance in comparison to other peaceful societies (Sullivan, Piereson, & Marcus, 1993), and increased intolerance towards gay and lesbian communities (UNAMI, 2006).

At a structural level, Rouhana and Bar-Tal (1998) have noted that in areas of intractable conflict, cultural and entertainment institutions such as theater, film, museums, and educational institutions all *align* together and further polarize societies. For example, educational institutions can fuel conflict by perpetuating societal beliefs and ideologies that instigate conflict (Bar-Tal, 1998) instead of keeping to their original mission of educating the general population (Bar-Tal, 1998; DeCharms & Moeller, 1962; Luke, 1988). This is achieved through the use of biased history and social studies textbooks as well as other curricular and pedagogical tools. In addition, we often observe a transformation in national history museums and art exhibits that propagate the "official" conflict narratives and often constitute collective memories for warring populations (Nets-Zehngut, 2008; Savelsberg & King, 2005).

An Exploratory Study on Pervasiveness

In 2009, an exploratory study was conducted in Israel-Palestine to investigate the phenomenon of conflict pervasiveness, with the goal of identifying its basic dimensions (Musallam et al., 2010). For the purposes of this study, conflict pervasiveness was defined for participants as *the range of issues and social processes that are organized into the structure of the conflict versus those that are "conflict free"*. The primary aim of the study was to try to distinguish those elements of day-to-day life in this setting that, at any given time, serve to drive the conflict (move people toward anger or aggression against outgroups) versus those that were perceived to be irrelevant to the conflict, or those elements that served to constrain the conflict (move people toward a sense of empathy or compassion with outgroups).

At the beginning of the study, participants were asked to think about a typical day in their-life and to list those activities that were "related to the conflict" and those that were "conflict free". Both Israelis and Palestinians found the task to be extremely challenging. For example, one of the Israeli participants argued: "*When you grow up in a conflict, conflict is your reality, that is life, that is the way things are. It is so difficult for me to distinguish what is related and what is not related to the conflict*". This experience was shared by others and illustrates how "conflict saturated" both Israeli and Palestinian societies are. When participants were presented with the statement: "*As the situation deteriorates and persists, more and more aspects of peoples' daily life become affected by it and serve to trigger or perpetuate a sense of anger/resentment/contempt*", they all agreed strongly with it.

All participants expressed strong emotional experiences that included a sense of helplessness, insecurity, hopelessness, frustration, anxiety, fear, and anger. For example, a sense of insecurity by Israeli participants reinforced the need to use violence against the Palestinians, whereas anger and hopelessness expressed by Palestinian participants supported the use of violent tactics against Israelis. In addition, some participants noted that as things escalated in their communities, the degree to which they participated in certain activities changed. For example, the number of times they checked news updates and the number of phone calls they made during particular days changed as a function of conflict escalation. As one of the participants put it "*It is not only about what you do, but how often you do certain things and for how long they last*".

Finally, both Israeli and Palestinian interviewees indicated that the degree to which they were affected by the conflict depended to a large extent on their *relative physical proximity* to high/low conflict intensity locations. Those residing closer to places where violence occurred experienced a much broader sense of contagion than those who lived in relatively more peaceful locations. However, Israelis viewed conflict pervasiveness as something that was more related to specific periods and incidents (e.g., when bombing attacks occurred), whereas Palestinians perceived conflict pervasiveness more consistently in their every-day experiences. In addition, for Israelis, "conflict pervasiveness" was associated mostly with psychological dimensions (e.g., decision making processes), whereas for Palestinians, conflict pervasiveness was associated with more tangible, mundane matters (e.g., the ability to go to a hospital, a store, etc.).

This exploratory study revealed several important points. First, it provided some preliminary empirical support of the subjective experience of the phenomenon of pervasiveness. Second, it highlighted the difficulty experienced by participants in distinguishing between what is related and what is unrelated to conflict in situations of protracted ethnopolitical conflict. Third, it pointed to the qualitative differences in experiences expressed by distinct populations. Finally, it highlighted the importance of understanding the various interrelated aspects of conflict pervasiveness, which include behavioral, cognitive, and affective components.

Although substantial amount of work has been dedicated to the development of measures for the assessment of conflict intensity, duration, and the many consequences of conflict intractability, to date no measures have been developed to assess their degree of pervasiveness. Developing such measures could have important implications for assessing long-term damage in conflict, diagnosing fragile and ripe peace processes (Coleman, 2000; Zartman, 1986, 2000, 2003), and assessing the effectiveness of interventions before applying them within conflicted societies (Coleman, 2006).

A Dynamical-Minimal Model of the Crude Law of Social Relations

Deutsch's Crude Law of Social Relations (1973) is a simple theoretical generalization of the results of a research program concerning the conditions that promote cooperation versus competition and the consequences of cooperating versus competing. It is based on the observation from empirical studies that the *consequences* of cooperation and competition are strikingly similar to their *causes*. In particular, the Crude Law states that "*the characteristic processes and effects elicited by a given type of social relationship (e.g., cooperative or competitive) tend also to elicit that type of social relationship; and a typical effect of any type of relationship tends to induce the other typical effects of the relationship*".

Basic Features of the Model

Following the approach of dynamical minimalism (Nowak, 2004), we attempted to model the Crude Law mathematically using the simplest possible assumptions (Nowak et al., 2010a). We adopted a generic formalism of growth—the *Alphabet Model* (Nowak & Solomon, 2006)—which is an elaboration of the AB model of Solomon and his collaborators (Shnerb, Louzoun, Bettelheim, & Solomon, 2000). The model concentrates on the development of protracted conflict across time.

Following the main factors constituting Crude Law dynamics as outlined by Deutsch (1973), the model is based on three types of elements:

- *Competitive interdependence*: The extent to which the interests of parties are incompatible.
- *Competitive behaviors*: The momentary behaviors of individuals to move against each other in order to obstruct the one another's goal achievement.
- *Competitive orientations*: The attitudes underlying competition. Such orientations have many components, such as low trust or negative perceptions of others. Orientations evolve on a slower time scale than do behaviors.

Computer Simulations of the Model

In our computational model, a 25 × 25 square grid represents the space where the conflict interactions occur. In the beginning of the simulation, a random number of competitive behaviors are assigned for each cell of the grid. Five cells (marked by the white dots in Fig. 5.2a) are randomly chosen as the places where competitive interdependence (e.g., competition over scarce resources such as water) exist. To model the progression of a conflict into intractability, the simulation starts with no elements containing competitive orientations. The starting configuration is shown in Fig. 5.2a.

The simulations are based on a few simple rules that operate locally (i.e., within a single cell). All the rules represent control parameters of the model and are applied with a specific probability, which can be adjusted. Thus, with a specific probability:

1. Competitive behaviors elicit new competitive behaviors.
2. Competitive behaviors decay with a certain rate.
3. Competitive behaviors migrate to nearby locations.
4. Competitive behaviors produce competitive orientations.
5. The probability of competitive behaviors producing new competitive behaviors grows with the local (same cell) presence of elements of competitive interdependence and also elements of competitive orientations.
6. With some small probability, each negative behavior migrates to a randomly chosen neighboring cell. Competitive behaviors thus diffuse from their place of origin.

Thus, in the presence of competitive interdependence, competitive behaviors reproduce competitive behaviors in kind and also help foster competitive orientations, which evolve at a slower rate. Competitive orientations also facilitate the multiplication of competitive behaviors. Each process thus produces conditions facilitating its occurrence at the place where the conditions are met—a process that captures the essence of the Crude Law.

The initial configuration (Fig. 5.2a), which is maintained in the early stages of the simulation (Fig. 5.2b), shows the initial random distribution of competitive interdependence and competitive behaviors. The height and color intensity of a box correspond to the number of competitive behaviors, with higher and darker colored

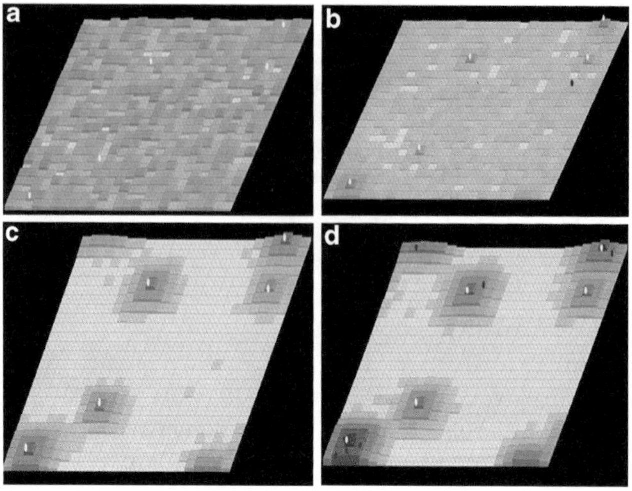

Fig. 5.2 (**a–d**) Simulation of the Crude Law

cells representing greater competition. As noted above, the small white dots correspond to the locations of competitive interdependence, or competing goals.

The initial dynamic is dominated by the decrease in the number of competitive behaviors. In the cells with no competitive interdependence, competitive behaviors vanish at a higher rate than such behaviors are produced. However, in the locations where structures of competitive interdependence are present, the number of negative behaviors grows as each competitive behavior produces another element of competitive behavior with a high probability. Because each competitive behavior most likely produces other competitive behaviors, the growth of competitive behaviors in the cells where structures of competitive interdependence exists is exponential. The number of competitive behaviors in the locations of competitive interdependence increases dramatically over time where there exists a longer history of interaction. Because the number of locations where the growth of conflict occurs is low, a general decrease in competitive behaviors can be observed.

Because of the exponential growth of competitive behaviors, after some time there is a global increase in competitive behaviors in the locations with high competitive interdependence (Fig. 5.2c). Due to the high number of competitive behaviors in the cells with competitive interdependence, competitive social orientations are eventually produced in these locations (Fig. 5.2d). Thus, islands of more destructive conflict characterized by competitive interdependence, behaviors, and orientations are established around these cells.

Figure 5.2c shows the formation of pools of destructive conflict. Figure 5.2d illustrates creation of competitive social orientations, which are marked by the dark dots. Note the high concentration of negative orientations in the lower left and upper right corner of Fig. 5.2d. The elements of competitive orientations were created in the locations where competitive interdependence existed. They were created by the high incidence of competitive behaviors, rather than directly by the competitive interdependence.

Reinstatement of Conflict

To simulate the effects of attempts at conflict resolution, in the next round of simulations the competitive behaviors were reset to the initial random configuration and the elements of competitive interdependence were eliminated. The final configuration of the previous simulation (Fig. 5.2d) was used as input. From a rational perspective, there is little reason for the conflict to exist any longer. The competitive orientations, however, were left intact. Figure 5.3a portrays the reinstatement of conflict in the area of high concentration of competitive orientations.

The final simulation was run for several steps. Figure 5.3b portrays the final configuration. Note the high concentration of competitive orientations in the lower left corner. Negative orientations act as a social memory of past conflict. The conflict is reinstated only in one of the original areas where the concentration of the competitive orientations was the highest. Through a reinforcing feedback loop, competitive behaviors produce more elements of competitive orientations. At this location, the conflict is *self-sustaining and growing at an accelerated rate*. However, in other areas where the conflict had existed previously, it was not reinstated. This is because the intensity of negative orientations elsewhere was not sufficient to sustain the process of competitive behavior growth.

The temporal evolution of the conflict, the intervention, and the reinstatement of the conflict are portrayed in Fig. 5.4 below. As we can see, the existence of competitive social orientations results in the conflict being reinstated almost immediately. Although the conflict was recreated in only one location, because of the fast exponential growth it takes only five simulation steps to reinstate a high level of competitive behaviors.

The results highlight two factors that are critical to escalation of conflicts: the duration of a conflict and the local severity of competitive interdependence. Simply put, a conflict will escalate the most if competitive interdependence exists in the same location and continues for a long time. If, however, the same amount of competitive interdependence ceases to exist in one place, and after a short time reappears in another, the severity of conflicts in a society will be much smaller. This is because in each location a conflict does not have enough time to change psychological orientations in a significant way, so it is less likely to escalate to intractability.

Thus, the model illustrates how conflicts tend to be recreated in places where they once existed, even after a long period of being dormant and even if competitive interdependence no longer exists. In other words, hostilities committed in a conflict create competitive orientations that reinstate the conflict in the same place.

This scenario suggests that the timing of intervention is critical for transforming conflicts. In the earlier stages of a conflict, intervention needs to be focused on competitive behaviors in the few places with conflicts of interest. In later stages,

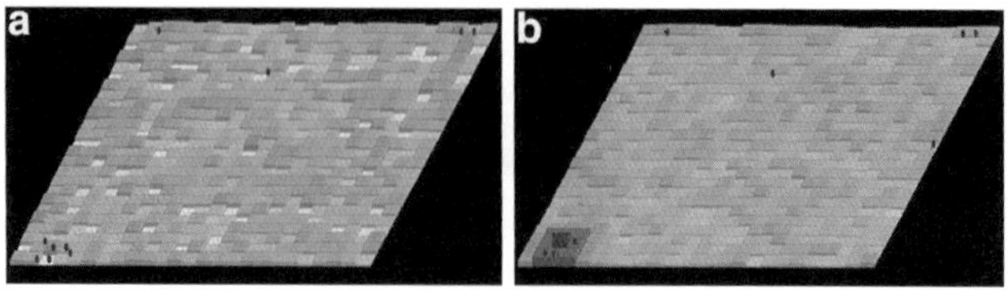

Fig. 5.3 (**a**, **b**) The initial and final configuration of the competitive orientations and competitive behaviors

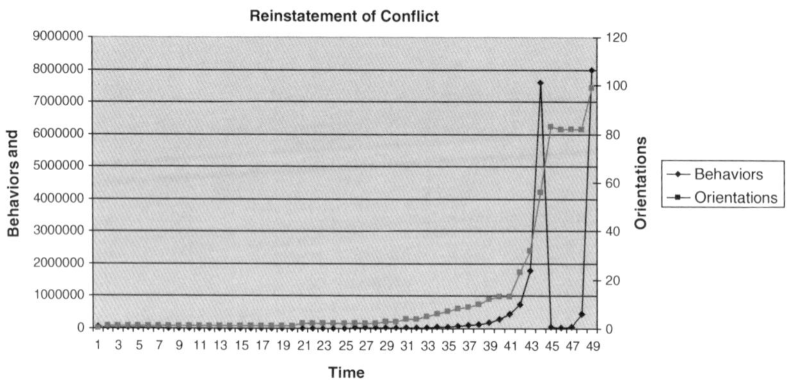

Fig. 5.4 Time course of the creation and reinstatement of the conflict

resolution of the conflict of interest still may be effective, although the conflict may have become global for the society and the levels of hostility are likely to be high. After some time, enough competitive orientations are created to sustain the conflict even in the absence of any conflict of interests. When the conflict becomes self-sustaining, it acquires the features of intractability.

Coping and Survival in Open Versus Closed Systems

From a dynamical systems perspective, it is important to distinguish between *closed* and *open* systems. In closed systems, dynamics depend entirely on the current state of the system (i.e., the current values of variables in the system) and are not influenced by input from the external environment. They may display emergent properties, but this tendency reflects only the self-organization of intrinsic dynamics without sensitivity to external factors that might promote the emergence of different properties. Open systems, in contrast, display dynamics that are responsive to environmental influences. Self-organization occurs in such systems, but some products of self-organization are better suited to environmental demands and are selected at the expense of less well-suited products of self-organization.

Not surprisingly, open systems have far greater capacity to cope and survive, particularly in a changing external environment. The notion of complex adaptive systems (e.g., Holland, 1975, 1995), for example, refers to the capacity of systems to change their own structures and functions in order to adapt to environmental factors. This is easy to appreciate in biological systems, which must be able to capitalize on opportunities such as the availability of food and the presence versus absence of predators. A plant that did not grow in the direction of sunlight, for example, would be at a severe disadvantage in the struggle for survival. Within the lifetime of a single organism, learning is the most efficient mechanism for adaptation. At the species level, evolution—based on mutation and natural selection—provides a mechanism for adaptation on a longer time scale that improves the fitness of successive generations of organisms. In short, systems that are closed cannot adapt and lose their fitness. Eventually, they are unable to cope and survive as the external environment changes.

The distinction between closed and open adaptive systems has important consequences for understanding intractable conflicts. Social groups, organizations, and societies that adapt to environmental influences have a clear advantage in coping and survival. In contrast, social systems that are inflexible in response to changing conditions are at risk for dysfunctional behavior and even extinction. The dynamical tendencies associated with intractable conflict capture the nature of closed systems. Indeed, the development of intractability may be described as the closing up of a previously open and adaptive system. Once the dynamics of a system are captured by a strong attractor, environmental events essentially represent perturbations that are actively resisted so as to maintain the system's coherent state (i.e., pattern of thoughts, feelings, and action tendencies). Thus, a social relation characterized by intractable conflict is composed of two (or more) parties—whether individuals, groups, or nations—that react towards one another with suspicion and hostility, even if such reactions are unwarranted, counterproductive, and maladaptive. Outside interventions, too, tend to be resisted in such systems, despite the long-term self-interest that may be associated with such interventions. Even after the issues that initially triggered the conflict have been resolved (or are no longer relevant), the parties to an intractable conflict remain intransient in their beliefs, emotions, and action orientations.

Intractable conflicts impair the capacity for coping and survival in other ways as well. As the conflict escalates and self-organizes into a stable attractor landscape, it enslaves an increasing number of psychological and social processes. Thus, many processes that usually serve basic needs of individuals and social groups lose their original functionality as they acquire new functionality of bringing harm to the opponent. As the system loses the functions necessary to fulfill its basic needs, it loses the capacity for coping and survival. This process is tantamount to the loss of complexity in system dynamics. Prior to the emergence of deep attractors for conflict, individuals and groups have rich, multidimensional dynamics enabling the achievement of various goals and the fulfillment of various functions. When the social relation devolves into a system of conflict, however, it absorbs other, previously functional processes, so that the dynamics of each party become organized along a single dimension of antagonism toward the other party.

This scenario can be seen at different levels of social reality. Consider marital conflict, for example. A married couple originally experiences rich dynamics organized along many dimensions, such as providing material support, caring for their children, organizing entertainment, cleaning their house, planning vacations, planning shopping, and organizing their social life. If a conflict develops that progresses toward intractability, however, these functions lose their independence and come to fuel rather than diffuse the conflict. Decisions concerning the children, for instance, may be governed by the desire to limit contact with the other parent rather than dictated by the child's needs. When planning entertainment, meanwhile, the motive may be to limit the pleasures of the partner rather than to spend enjoyable time together. And cleaning the house may become an occasion to blame the partner for the mess and assigning motives and personality traits that justify the judgment. As an increasing number of activities become transformed and aimed at harming the partner, the original function of these activities are lost. As a result, the capacity of the couple to cope with the flux of life is likely to be dramatically reduced.

International conflict that devolves into intractability can be seen in the same light. The large losses experience by warring nations to a large extent reflect a loss of functionality in each nation's social processes and institutions. As a conflict escalates and becomes intractable, the scope of social processes and institutions relevant to the conflict increases. The educational system, for example, becomes a system of propaganda rather than one of teaching children. The priority of transportation changes from the delivery or food and supplies to the movement of soldiers and weapons. The function of the justice system changes from enforcing laws to prosecuting (and persecuting) enemies in the outgroup and members of the ingroup who are suspected of supporting the outgroup enemies. The main function of the economic system is altered to finance the machinery of war, rather than to provide the material well being of the society. As these different institutions and processes acquire new (dis)functionality related to conflict, they lose their original purpose and thus the society becomes less able to fulfill the needs of its citizens. In general, all social functions suffer as a society loses its ability to cope with the variety of social and economic challenges it faces.

Yet another maladaptive consequence of intractable conflict concerns the temporal focus of the parties to the conflict. Effective adaptation requires that systems react to present events in the current state of their environment. When locked in intractable conflict, however, the parties tend often to focus on the past (e.g., past injustices and grievances, seminal events that are memoralized, the history of the relationship) rather than on the needs of the present. Fixation on the past decouples the system from its focus on the present, so that the respective parties react to events that are no longer current and cannot be changed. And when the parties fixate on the past, they are likely to ignore signals that the opposing party is ready for reconciliation, and may misinterpret peaceful overtures. This clearly limits the prospect for conflict transformation. Politicians during times of conflict often exploit the tendency to fixate on the past so as to prevent reconciliation that may not serve their personal ambitions and interests. Playing on people's fears and hatred is a good way to secure and maintain power in a time of crisis and strained international relations.

So, What Can Be Done?

In our account, intractable conflict has all the hallmarks of a system that has become self-organized with respect to a strong attractor that provides a coherent way of thinking, feeling, and acting. As is the case for any dynamical system, once a deep attractor for conflict emerges in a social system, it is highly resistant to change. Rather than accommodating to new information and being sensitive to changing challenges and demands, a social system governed by a strong attractor reacts to such influences as if they were perturbations that threaten the system's integrity. A host of cognitive and social mechanisms are employed to deflect or reinterpret such perturbations, rendering the system incapable of responding appropriately and rationally to changing circumstances or even acting in its own best long-term interests. A relationship mired in intractable conflict is inflexible and closed—hardly the characteristics of an adaptive system.

The dynamical scenario of intractable conflict seems gloomy, if not hopeless. How can one possibly transform a system that has lost its complexity and openness to change, and whose sole motive is the preservation of a fixed way of seeing the world? Actually, this is far from the end of the story. Indeed, although it may seem ironic, the strategies that may prove effective in transforming intractable conflict follow from the principles of dynamical systems theory and research. This aspect of dynamical systems—their potential for change in their basic processes—is the focus of Chap. 6.

References

Aron, A., & Corne, S. (1994). *Writings for a liberation of psychology*. Boston, MA: Harvard University Press.

Azar, E. E. (1983). The theory of protracted social conflict and the challenge of transforming conflict situations. *Monograph Series in World Affairs, 20*, 81–99.

Azar, E. E. (1990). *The management of protracted social conflict*. Hampshire, UK: Dartmouth.

Baker-Brown, G., Ballard, E. J., Bluck, S., deVries, B., Suedfeld, P., & Tetlock, P. (1992) The integrative complexity coding manual. In C. Smith (Ed.), *Handbook of thematic analysis* (pp. 605–611). Cambridge, England: Cambridge University Press.

Bales, R. F. (1950). *Interaction process analysis*. Chicago, IL: University of Chicago Press.

Bar-Tal, D. (1998). Societal beliefs in times of intractable conflict: The Israeli case. *International Journal of Conflict Management, 9*, 22–50.

Bar-Tal, D., Raviv, A., & Freund, T. (1994). An anatomy of political beliefs: A study of their centrality, contents, and epistemic authority. *Journal of Applied Social Psychology, 24*, 849–872.

Baumeister, R. F., Bratslavsky, E., Muraven, M., & Tice, D. M. (1998). Ego-depletion: Is the active self a limited resource? *Journal of Personality and Social Psychology, 74*, 1252–1265.

Baumeister, R. F., & Heatherton, T. F. (1996). Self-regulation failure: An overview. *Psychological Inquiry, 7*, 1–15.

Baumeister, R. F., Vohs, K. D., & Tice, D. M. (2007). The strength model of self-control. *Current Directions in Psychological Science, 16*, 351–355.

Burgess, H., & Burgess, G. (1996). Constructive confrontation: A transformative approach to intractable conflicts. *Mediation Quarterly, 13*, 305–322.

Carnevale, P. J., & Probst, T. (1998). Social values and social conflict in creative problem solving and categorization. *Journal of Personality and Social Psychology, 74*, 1300–1309.

Carver, C. S., & Scheier, M. F. (1999). Stress, coping and self-regulatory processes. In L. Pervin & J. P. Oliver (Eds.), *Handbook of personality: Theory and research* (pp. 553–575). New York, NY: Guilford.

Coleman, P. T. (2000). Fostering ripeness in seemingly intractable conflict: An experimental study. *International Journal of Conflict Management, 11*, 300–317.

Coleman, P. T. (2003). Characteristics of protracted intractable conflict. *Peace and Conflict: Journal of Peace Psychology, 9*, 1–37.

Coleman, P. T. (2004). Paradigmatic framing of protracted, intractable conflict: Towards the development of a meta-framework—II. *Peace and Conflict: Journal of Peace Psychology, 10*, 197–235.

Coleman, P. T. (2006). Conflict, complexity, and change: A meta-framework for addressing protracted, intractable conflicts - III. *Peace and Conflict: Journal of Peace Psychology, 12*, 325–348.

Coleman, P. T. (2011). *The five percent: Finding solutions to (seemingly) impossible conflicts*. New York, NY: Public Affairs, Perseus Books.

Coleman, P. T., Bui-Wrzosinska, L., Vallacher, R. R., & Nowak, A. (2006). Protracted conflicts as dynamical systems. In A. K. Schneider & C. Honeyman (Eds.), *The negotiator's fieldbook: The desk reference for the experienced negotiator* (pp. 61–74). Chicago, IL: American Bar Association Books.

Coleman, P. T., Vallacher, R. R., Nowak, A., & Bui-Wrzosinska, L. (2007). Intractable conflict as an attractor: Presenting a model of conflict, escalation, and intractability. *The American Behavioral Scientist, 50*, 1454–1475.

Comas-Díaz, L., Lykes, M. B., & Alarcón, R. (1998). Ethnic conflict and psychology of liberation in Guatemala, Perú, and Puerto Rico. *The American Psychologist, 53*, 778–792.

Crocker, C. A., & Hampson, F. O. (2004). *Taming intractable conflicts: Mediation in the hardest cases*. Washington, DC: United States Institute of Peace.

Crocker, C. A., Hampson, F. O., & Aall, P. (1999). *Herding cats: Multiparty mediation in a complex world*. Washington, DC: United States Institute of Peace.

Crocker, C. A., Hampson, F. O., & Aall, P. (2005). *Grasping the nettle: Analyzing cases of intractable conflict*. Washington, DC: United States Institute of Peace.

DeCharms, R., & Moeller, G. H. (1962). Values expressed in American children's readers. *Journal of Abnormal and Social Psychology, 64*, 136–142.

Dechesne, M., Greenberg, J., Arndt, J., & Schimel, J. (2000). Terror management and the vicissitudes of sports fan affiliation: The effects of mortality salience on optimism and fan identification. *European Journal of Social Psychology, 30*, 813–835.

Deutsch, M. (1973). *The resolution of conflict: Constructive and destructive processes*. New Haven, CT: Yale University Press.

Diamond, L., & McDonald, J. (1996). *Multi-track diplomacy: A systems approach to peace*. West Hartford, CT: Kumarian Press.

Druckman, D. (2001). Turning points in international negotiation: A comparative analysis. *Journal of Conflict Resolution, 45*, 519–544.

Fredrickson, B. L. (1998). What good are positive emotions? *Review of General Psychology, 2*, 300–319.

Fredrickson, B. L. (2001). The role of positive emotions in positive psychology: The broaden-and-build theory of positive emotions. *The American Psychologist, 56*, 218–226.

Fredrickson, B. L. (2006). Unpacking positive emotions: Investigating the seeds of human flourishing. *The Journal of Positive Psychology, 1*, 57–59.

Gailliot, M. T., Baumeister, R. F., DeWall, C. N., Maner, J. K., Plant, E. A., Tice, D. M., et al. (2007). Self-control relies on glucose as a limited energy source: Willpower is more than a metaphor. *Journal of Personality and Social Psychology, 92*, 325–336.

Gilbert, D. T. (1989). Thinking lightly about others: Automatic components of the social inference process. In J. S. Uleman & J. A. Bargh (Eds.), *Unintended thought* (pp. 189–211). New York, NY: Guilford Press.

Gilbert, D. T. (1993). The assent of man: Mental representation and the control of belief. In D. M. Wegner & J. W. Pennebaker (Eds.), *The handbook of mental control* (pp. 57–87). Englewood Cliffs, NJ: Prentice Hall.

Gilbert, D. T., & Hixon, J. G. (1991). The trouble of thinking: Activation and application of stereotypic beliefs. *Journal of Personality and Social Psychology, 60*, 509–517.

Gilbert, D. T., Pelham, B. W., & Krull, D. S. (1988). On cognitive busyness: When person perceivers meet persons perceived. *Journal of Personality and Social Psychology, 54*, 733–739.

Glass, D. C., & Singer, J. E. (1972). Behavioral aftereffects of unpredictable and uncontrollable aversive events. *American Scientist, 60*, 457–465.

Gottman, J. (2002). A multidimensional approach to couples. In *Comprehensive handbook of psychotherapy: Cognitive behavioral approaches* (Vol. 2, pp. 355–372). New York, NY: Wiley.

Gottman, J. M., Markman, H., & Notarius, C. (1977). The topography of marital conflict: A sequential analysis of verbal and nonverbal behavior. *Journal of Marriage and the Family, 39*, 461–477.

Gottman, J. M., & Silver, N. (1999). *The seven principles for making marriage work.* New York, NY: Three Rivers Press.

Gottman, J. M., Swanson, C., & Swanson, K. (2002). A general systems theory of marriage: Nonlinear difference equation modeling of marital interaction. *Personality and Social Psychology Review, 6*, 326–340.

Gray, B., Coleman, P. T., & Putnam, L. L. (2007). Introduction: Intractable conflict—New perspectives on the causes and conditions for change. *The American Behavioral Scientist, 50*, 1415–1429.

Greenberg, J., Porteus, J., Simon, L., Pyszczynski, T., & Solomon, S. (1995). Evidence of a terror management function of cultural icons: The effects of mortality salience on the inappropriate use of cherished cultural symbols. *Personality and Social Psychology Bulletin, 21*, 1221–1228.

Greenberg, J., Pyszczynski, T., Solomon, S., Rosenblatt, A., Veeder, M., Kirkland, S., et al. (1990). Evidence for terror management theory II: The effects of mortality salience on reactions to those who threaten or bolster the cultural worldview. *Journal of Personality and Social Psychology, 58*, 308–318.

Holland, J. (1975). *Adaptation in natural and artificial systems.* Ann Arbor, MI: The University of Michigan Press.

Holland, J. (1995). Can there be a unified theory of complex adaptive systems? In H. J. Morowitz & J. L. Singer (Eds.), *The mind, the brain, and complex adaptive systems.* Reading, MA: Addison-Wesley.

Johnson, D. W., & Johnson, R. T. (2006). Peace education for consensual peace: The essential role of conflict resolution. *Journal of Peace Education, 3*, 147–174.

Kelman, H. C. (1997). Social-psychological dimensions of international conflict. In I. W. Zartman & J. L. Rasmussen (Eds.), *Peacemaking in international conflict: Methods and techniques* (pp. 191–237). Washington, DC: United States Institute of Peace Press.

Kriesberg, L. (2005). Nature, dynamics, and phases of intractability. In C. A. Crocker, F. O. Hampson, & P. Aall (Eds.), *Grasping the nettle: Analyzing cases of intractable conflict* (pp. 65–98). Washington, DC: United States Institute of Peace.

Kruglanski, A. W., & Webster, D. M. (1996). Motivated closing of the mind: "Seizing" and "freezing. *Psychological Review, 103*, 263–283.

Kugler, K. G., Coleman, P. T., & Fuchs, A. M. (2011). Conflict, Complexity, and Openness: Constructive versus Destructive Dynamics of Discussions over Intractable Issues. *WOP Working Papers No. 2011/3*. Retrieved from: www.psy.lmu.de/wirtschaftspsychologie/forschung/working_papers/index.html.

Kugler, K. G., & Brodbeck, F. C. (2011). *How organizational communication at the top relates to conflict management and organizational justice perceptions on the work floor*. In WOP working paper no. 2011/1. Retrieved from http://www.psy.lmu.de/wirtschaftspsychologie/forschung/working_papers/index.html.

Latané, B., & Nowak, A. (1994). Attitudes as catastrophes: From dimensions to categories with increasing involvement. In R. R. Vallacher & A. Nowak (Eds.), *Dynamical systems in social psychology* (pp. 219–249). San Diego, CA: Academic.

Losada, M. (1999). The complex dynamics of high performance teams. *Mathematical and Computer Modelling, 30*, 179–192.

Losada, M., & Heaphy, E. (2004). The role of positivity and connectivity in the performance of business teams: A nonlinear dynamics model. *The American Behavioral Scientist, 47*, 740–765.

Luke, A. (1988). *Literacy, textbooks and ideology*. London: Falmer.

McClelland, J. L., Rumelhart, D. E., & Hinton, G. E. (1986). *Parallel distributed processing: Explorations in the microstructure of cognition* (Psychological and biological models, Vol. 2). Cambridge, MA: MIT Press.

Miller, G. A., Galanter, E., & Pribram, K. (1960). *Plans and the structure of behavior*. New York, NY: Holt.

Morris, M., & Kretzchmar, M. (1995). Concurrent partnerships and transmission dynamics in networks. *Social Networks, 17*, 299–318.

Musallam, N., Coleman, P. T., & Nowak, A. (2010). Understanding the spread of malignant conflict: A dynamical systems approach. Peace and Conflict. *The Journal of Peace Psychology, 16*, 127–151.

Nets-Zehngut, R. (2008). The Israeli national information center and collective memory of the Israeli–Arab conflict. *The Middle East Journal, 62*, 653–670.

Nowak, A. (2004). Dynamical minimalism: Why less is more in psychology? *Personality and Social Psychology Review, 8*, 183–193.

Nowak, A., Deutsch, M., Bartkowski, W., & Solomon, S. (2010a). From crude law to civil relations: The dynamics and potential resolution of intractable conflict. *Peace and Conflict: Journal of Peace Psychology, 16*, 189–209.

Nowak, A., & Solomon, S. (2006). Alphabet model of social processes. Paper presented at the conference "*Complexity models of social and economic processes*" Torino, Italy.

Nowak, A., & Vallacher, R. (1998). *Dynamical social psychology*. New York, NY: Guilford Press.

Nowak, A., Vallacher, R. R., Bui-Wrzosinska, L., & Coleman, P. T. (2007). Attracted to conflict: A dynamical perspective on malignant social relations. In A. Golec de Zavala & K. Skarzynska (Eds.), *Understanding social change: Political psychology in Poland* (pp. 33–49). Hauppauge, NY: Nova Science.

Nowak, A., Vallacher, R. R., Tesser, A., & Borkowski, W. (2000). Society of self: The emergence of collective properties in self-structure. *Psychological Review, 102*, 39–61.

Operario, D., & Fiske, S. T. (2001). Ethnic identity moderates perceptions of prejudice: Judgments of personal versus group discrimination and subtle versus blatant bias. *Personality and Social Psychology Bulletin, 27*, 550–561.

Powers, W. T. (1973). *Behavior: The control of perception*. Chicago, IL: Aldine.

Read, S. J., & Miller, L. C. (2002). Virtual personalities: A neural network model of personality. *Personality and Social Psychology Review, 6*, 357–369.

Read, S. J., Vanman, E. J., & Miller, L. C. (1997). Connectionism, parallel constraint satisfaction processes, and Gestalt principles: (Re)introducing cognitive dynamics to social psychology. *Personality and Social Psychology Review, 1*, 26–53.

Rosenblatt, A., Greenberg, J., Solomon, S., Pyszczynski, T., & Lyon, D. (1989). Evidence for terror management theory: I. The effects of mortality salience on reactions to those who violate or uphold cultural values. *Journal of Personality and Social Psychology, 57*, 681–690.

Rouhana, N. N., & Bar-Tal, D. (1998). Psychological dynamics of intractable ethnonational conflicts: The Israeli–Palestinian case. *The American Psychologist, 53*, 761–770.

Satish, U. (1997). Behavioral complexity: A review. *Journal of Applied Social Psychology, 27*, 2047–2067.

Saunders, A. D. (1997). *Channel defenses*. London: BT Batsford.

Savelsberg, J., & King, R. D. (2005). Institutionalizing collective memories of hate: Law and law enforcement in Germany and the United States. *The American Journal of Sociology, 111*, 579–616.

Shamir, M., & Sagiv-Schifter, T. (2006). Conflict, identity, and tolerance: Israel in the Al-Aqsa Intifada. *Political Psychology, 27*, 569–595.

Shnerb, N. M., Louzoun, Y., Bettelheim, E., & Solomon, S. (2000). The importance of being discrete: Life always wins on the surface. *Proceedings of the National Academy of Science, 97*, 10322.

Shoda, Y., Tiernan, L., & Mischel, W. (2002). Personality as a dynamical system: Emergence of stability and distinctiveness from intra- and interpersonal interactions. *Personality and Social Psychology Review, 6*, 316–325.

Staub, E. (2001). Genocide and mass killing: Their roots and prevention. In D. J. Christie, R. V. Wagner, & D. D. Winter (Eds.), *Peace, conflict and violence: Peace psychology for the 21st century* (pp. 76–86). Upper Saddle River, NJ: Prentice Hall.

Sullivan, J. L., Piereson, J., & Marcus, G. E. (1993). *Political tolerance and American democracy*. Chicago, IL: University of Chicago Press.

Sullivan, J. L., Shamir, M., Walsh, P., & Roberts, N. S. (1985). *Political tolerance in context: Support for un-popular minorities in Israel New Zealand, and the United States*. Boulder, CO: Westview.

Tesser, A. (1978). Self-generated attitude change. *Advances in Experimental Social Psychology, 1*, 289–338.

Tice, D. M., Bratslavsky, E., & Baumeister, R. F. (2001). Emotional distress regulation takes precedence over impulse control: If you feel bad, do it! *Journal of Personality and Social Psychology, 80*, 53–67.

UNFPA. (2006). *UNFPA state of the world population. United Nations population fund*. New York, NY: United Nations, Department of Public Information.

United Nations Assistance Mission for Iraq (UNAMI). (2006). *Human rights report 1 May—30 June 2006*. New York, NY: UNAMI.

Vallacher, R. R., Coleman, P. T., Nowak, A., & Bui-Wrzosinska, L. (2010). Rethinking intractable conflict: The perspective of dynamical systems. *The American Psychologist, 65*(4), 262–278.

Vallacher, R. R., & Nowak, A. (1999). The dynamics of self-regulation. In R. S. Wyer Jr. (Ed.), *Advances in social cognition* (Vol. 12, pp. 241–259). Mahwah, NJ: Lawrence Erlbaum.

Vallacher, R. R., & Nowak, A. (2007). Dynamical social psychology: Finding order in the flow of human experience. In A. W. Kruglanski & E. T. Higgins (Eds.), *Social psychology: Handbook of basic principles* (2nd ed., pp. 734–758). New York, NY: Guilford Press.

Vallacher, R. R., Nowak, A., & Kaufman, J. (1994). Intrinsic dynamics of social judgment. *Journal of Personality and Social Psychology, 67*, 20–34.

Vallacher, R. R., & Wegner, D. M. (1987). What do people think they're doing? Action identification and human behavior. *Psychological Review, 94*, 3–15.

Vohs, K. D., Baumeister, R. F., & Ciarocco, N. J. (2005). Self-regulation and self-presentation: Regulatory resource depletion impairs impression management and effortful self-presentation depletes regulatory resources. *Journal of Personality and Social Psychology, 88*, 632–657.

Vohs, K. D., Baumeister, R. F., Schmeichel, B. J., Twenge, J. M., Tice, D. M., & Nelson, N. M. (2008). Making choices impairs subsequent self-control: A limited resource account of decision making, self-regulation, and active initiative. *Journal of Personality and Social Psychology, 94*, 883–898.

Vohs, K. D., & Heatherton, T. F. (2000). Self-regulatory failure: A resource-depletion approach. *Psychological Science, 11*, 249–254.

Wegner, D. M. (1994). Ironic processes of mental control. *Psychological Review, 101*, 34–52.

Wegner, D. M., Erber, R., & Zanakos, S. (1993). Ironic processes in the mental control of mood and mood-related thought. *Journal of Personality and Social Psychology, 65*, 1093–1104.

Wegner, D. M., & Pennebaker, J. W. (Eds.). (1993). *Handbook of mental control*. Englewood Cliffs, NJ: Prentice-Hall.

Wegner, D. M., Schneider, D., Carter, S. R., & White, T. L. (1987). Paradoxical effects of thought suppression. *Journal of Personality and Social Psychology, 53*, 5–13.

Weiner, N. (1948). *Cybernetics, or control and communication in the animal and the machine*. Cambridge, MA: MIT Press.

Wessells, M. G. (2007). Post-conflict healing and reconstructing for peace: The power of social mobilization. In J. D. White & A. J. Marsella (Eds.), *Fear of persecution: Global human rights, international law and human well being* (pp. 258–278). Lanham, MD: Lexington Books.

Zartman, I. W. (1986). Ripening conflict, ripe moment, formula and mediation. In D. BenDahmane & J. McDonald (Eds.), *Perspectives on negotiation*. Washington, DC: GPO.

Zartman, I. W. (1989). *Ripe for resolution: Conflict and intervention in Africa* (2nd ed.). New York, NY: Oxford University Press.

Zartman, I. W. (2000). Ripeness: The hurting stalemate and beyond. In P. C. Stem & D. Druckman (Eds.), *International conflict resolution after the cold war* (pp. 225–250). Washington, DC: National Academy Press.

Zartman, I. W. (2003). Negotiating with terrorists. *International Negotiation, 8*, 443–450.

Zartman, W. A. (2005). Rethinking the economics of war. In C. J. Arnson (Ed.), *The intersection of need, creed, and greed*. Baltimore, MD: Johns Hopkins University Press.

Chapter 6
Escape: How Conflicts Can Be Transformed

Kurt Lewin (1948) famously observed, "there is nothing so practical as a good theory." This simple statement captures a truism regarding the interplay of understanding, prediction, and control that characterizes every area of science. Predicting how a phenomenon will be manifest under different conditions, let alone controlling the process, is intimately linked to a coherent and generalized understanding of the phenomenon at issue. Humans, after all, did not land on the moon or send satellites to other planets by focusing on how to do these things. Space exploration would have remained a flight of fancy had it not been for several centuries of scientific concern with basic principles of physics and chemistry. This realization is relevant to the understandable concern people have for resolving the difficult and protracted conflicts that characterize interpersonal, inter-group, and international relations in today's world. Practitioners are motivated to tackle such conflicts head-on, but their likelihood of success is ultimately constrained by the degree of scientific understanding concerning far more basic and mundane aspects of psychology.

Accordingly, the preceding chapters have focused on identifying basic dynamic processes that underlie human experience at different levels of personal, interpersonal, and collective functioning. Indeed, the perspective we have presented is sufficiently general that it characterizes phenomena of all kinds in physical as well as social reality. With this comprehensive understanding in mind, the aim of this chapter is to put Lewin's dictum to the test: Can effective strategies of intervention be developed and implemented that reflect the adaptation of dynamical principles to intractable human conflict?

To set the stage for this translation of theory into practice, we discuss the difficulties practitioners have experienced in their attempts to resolve difficult conflicts. Many intuitively reasonable strategies, some of them deeply insightful, have come up empty in transforming hostile and intransient relations into benign or constructive relations. These disappointments are actually quite informative—and understandable from the dynamical perspective. We then reframe conflict resolution in terms of attractor dynamics and offer three basic ways in which this understanding can be implemented to resolve conflicts that appear intractable.

R.R. Vallacher et al., *Attracted to Conflict: Dynamic Foundations of Destructive Social Relations*, Peace Psychology Book Series, DOI 10.1007/978-3-642-35280-5_6,

The chapter concludes by noting two recent efforts to translate these basic scenarios into training programs for negotiators and others concerned with difficult conflicts. The interested reader is directed to Appendices A and B to learn the specifics of these programs.

The Uphill Battle in Conflict Resolution

One might think that resolving a conflict should be a relatively straightforward matter. At its base, conflict centers on the perception of incompatible needs or interests (cf. Deutsch, Coleman, & Marcus, 2006). The task, from this perspective, is how to eliminate or minimize the perceived incompatibility—or even better, how to foster common goals, interests, and values. Even when the incompatibility cannot be totally eliminated, it should be possible to fashion a compromise that partially addresses the most pressing needs and interests of the parties. This seems especially reasonable for a conflict that has become protracted and destructive. A conflict with no end in sight, after all, does little to satisfy the interests of either party, and it represents a huge drain on both parties' resources, energy, and human capital in service of what appears to be a futile endeavor. A compromise may not be ideal for either party, but it represents a solution that is clearly preferable to continuing to pursue a course of action that promises nothing but aversive outcomes and little likelihood of goal attainment.

In fact, most conflicts are resolved through compromise and negotiation. But in a small percentage of conflicts—about 5–8 %—the parties defy common sense and logic, and become locked into an antagonistic state that seems impervious to resolution (Coleman, 2011; Deihl & Goertz, 2001). Indeed, even after the perceived incompatibilities that launched the conflict are no longer at issue, the parties to an intractable conflict maintain their mutually destructive relationship with no end in sight (e.g., Azar, 1990; Bennett, 1996; Bercovitch, 2005; Burton, 1987; Coleman, 2003; Goertz & Diehl, 1993; Kriesberg, 2005; Marshall & Gurr, 2005; Pearce & Littlejohn, 1997). Faced with such intransigence and irrationality, those who attempt to resolve intractable conflicts have an uphill battle on a number of fronts. To deal with these challenges, practitioners need to rethink some assumptions that, although intuitively reasonable, do not square with the dynamical nature of intractable conflict. Below are ten such assumptions (see Coleman, 2011, for fuller elaboration).

Comparing Fluid Things to Fixed Things

Psychologists and social scientists tend to look for stable qualities in people and social systems, and then investigate how these qualities are expressed in various contexts. Individuals are presumed to have stable personality traits (e.g., Costa &

McCrae, 1995), for example, and cultures are differentiated in terms of basic values and social orientations such as independence versus interdependence (e.g., Triandis, 1995). If these qualities can be specified, they should enable theorists and practitioners to predict—and ultimately, control—how individuals and cultures will behave in a scenario of conflict. The problem is that individuals and cultures are not static entities with fixed ways of thinking and acting, but rather they demonstrate considerable fluidity in their mental and behavioral orientations. The most agreeable or intellectually curious person can become hostile and close-minded when his or her deeply held beliefs are challenged, for example, and people in a highly independent culture can display impressive group solidarity when faced with an external threat.

Research that assumes stability rather than variability in people is likely to come up empty when trying to understand, predict, and control a phenomenon as multi-determined and volatile as conflict. Efforts to link static personality variables and *a priori* cultural differences to patterns of conflict and the success of conflict resolution strategies have produced largely inconsistent and confusing results (Lewicki, Literer, Minton, & Saundars, 2004; Zartman & Faure, 2005). Variability and the potential for change in system elements and global system properties go to the heart of the dynamical perspective. But beyond simply acknowledging this fact of human experience, theory and research informed by complexity science and dynamical systems specify the nature of such changes and under what conditions they can be expected to occur and play a substantial role in the generation, maintenance, and resolution of conflict.

Thinking in Straight Lines

Linear causality, with events unfolding in accordance with a sequence of well-defined individual and isolated cause and effect relations, seems like a reasonable model of the world and has been a cornerstone of science since its beginnings. In the context of theory and research on social conflict, this model suggests that a particular set of conditions (e.g., external threat, resource scarcity) will invariably lead to a particular pattern of behavior (e.g., defiance, competition). Simple cause-effect relations can certainly be observed, but they are embedded in a larger field of forces that can magnify, nullify, or even reverse these relations. Considered in isolation, for example, an external threat might indeed produce defiance and resource scarcity might elicit competition between individuals, groups, or nations. But such conditions are rarely—probably never—experienced apart from a host of other conditions that interact with the "causal factor" under investigation. How a person or social system responds to a threat from another entity, for example, depends on such things as the relative power of the respective entities, their prior history, and their relative support each receives from other entities. The same can be said for the effect of scarce resources. Under certain patterns of conditions and forces, in fact, a scarce resource can promote cooperation rather than competition (e.g., Axelrod, Riolo, & Cohen, 2002).

By their very nature, complex dynamical systems call into question the utility of focusing on simple cause-effect relations that are decoupled from the larger context in which they occur. Experiments are useful, of course, in exposing plausible causal scenarios, but ultimately these scenarios play out in settings that reflect myriad other causal scenarios, many of which are mutually interdependent. It is also the case that causality in dynamical systems is often bi-directional, so that the effect observed at one point in time may function as a cause at the next point in time. If an external threat promotes anger and an aggressive pre-emptive response, for example, the threatening party may double-down on its threatening behavior—or back off entirely—depending on a host of other factors. Clearly, then, deep understanding of a phenomenon as complex as conflict requires a multi-disciplinary strategy, with feedback among experiment, computer simulation, field studies, archival data, and historical records—a point we have stressed in earlier chapters.

Privileging of the Short-Term

One can imagine a conflict that ends as quickly as it begins. Two school children race to a drinking fountain after a sweat-inducing recess between classes, for example, and one of them wins the race and drinks while the other waits. Quite likely, that would pretty much put an end to the conflict. In this case, one need only focus on the immediate effect and not consider what happens later (unless the losing child holds a grudge or the winning child sees the relationship from then on in winner-loser terms). Most conflicts do not conform to this scenario, but rather unfold over time. In the case of intractable conflicts, of course, this can be a considerable period of time. The short-term effects in such conflicts may not be predictive of what happens in the long-term. A group's immediate reaction to an external threat might be anger and defiance, for example, but over time the reaction might reverse itself, with a desire for compromise or perhaps outright submission becoming salient. The same concern can be voiced with respect to conflict resolution strategies. A strategy might have the immediate effect of lowering the intensity of a conflict, only to produce a substantial rebound in antagonism if the initial expectations are not fulfilled. By the same token, a strategy that is rejected at the outset may become appreciated and accepted after a period of time.

Temporal patterns are front and center in the investigations of dynamical systems, particularly those systems characterized by nonlinear relations among variables (e.g., Holland, 1995; Kelso, 1995; Nowak & Vallacher, 1998; Schuster, 1984). As noted in Chap. 3, there are a handful of basic dynamic patterns, and each is informative about the structure of underlying forces in the system. In investigating conflict, then, it is not sufficient to look at the long-term effects. Instead, one should focus on the form of the conflict trajectory. The immediate impact of a factor or set of factors may be magnified or diminished over time, for example, or it could promote a complex pattern of variability, such as periodic evolution or chaos. Attention to this critical feature of nonlinear dynamical systems is essential to the understanding, prediction, and possible control of a conflict scenario.

Focusing on Problems

It may seem self-evident that conflict and peace exist as endpoints on a common dimension. Conflict exists to the extent that peaceful relations are non-existent, and peace exists when conflict is minimized or resolved. In studying conflict, then, there is no need to focus on peace as an independent condition—to do so would be redundant. This assumption actually reflects a larger perspective in psychology that focuses on the problematic features of thought, emotion, and action. In this view, health and adaptation is the absence of problems, so that eliminating one state (e.g., distrust in a relationship) will bring about the opposite state (e.g., trust). Recent advances in psychology, however, have challenged this assumption and focused the attention of theorists and researchers on the possibility that seemingly opposite states actually represent independent dimensions. In the study of emotion, for example, positive and negative affective states can be experienced in various combinations (e.g., Cacioppo & Bernston, 1994). A person may have very positive feelings regarding a topic, but this does not rule out the possibility that he or she has equally strong negative feelings toward the topic. Close relationships, too, can be characterized by simultaneous feelings of love and hate, cooperation and resentment, empathy and judgment (e.g., Gottman, Swanson, & Swanson, 2002).

This has important implications for the understanding, prediction, and control of conflicts. The relations between individuals, groups, or societies can reflect different blends of seemingly incompatible feelings and action orientations. Two countries may have propensities for both mutual respect and mutual resentment, warm relations and overt hostility, trust and suspicion. Focusing only on reducing the problematic dimension—resentment, hostility, or suspicion—does not guarantee that the other relationship quality—respect, warmth, or trust—will show a proportional increase. It is understandable that those concerned with resolving conflicts would focus on the negative and destructive aspects of a relationship. But without attending to the potential for positivity in the relationships, an effective and sustained solution may prove elusive. The importance of focusing on positive and constructive qualities rather than on faulty or unwanted system functioning is also a primary lesson of the emerging field of positive psychology (e.g., Seligman, Steen, & Park, 2005).

The realization that conflict and peace are separate dimensions and should be treated as such follows directly from the potential for multi-stability in dynamical systems. Indeed, as we noted in Chap. 3, the very existence of one coherent state (e.g., negative feelings, competitive orientation) is the precondition for the emergence of the opposite state (positive feelings, cooperative orientation). This is because once a system converges on an attractor, new events and information that cannot be assimilated to that attractor become self-organized with respect to an alternative coherent state that can exist as a latent attractor, ready to take control of the system's dynamics under certain conditions. From a dynamical point of view, then, the problem of malignant relations between individuals, groups, or societies is not solved when a conflict is resolved. One must also focus on the dynamics of peace (cf. Nan, Mampilly, & Bartoli, 2012).

Marginalizing Emotions

When we see two people exhibiting intense emotion in a heated confrontation, we are tempted to advise them to calm down so that the dispute can be handled in a rational manner. Strong emotions, after all, get in the way of clear thinking and allow the conflict to get blown way out of proportion. Much the same view predominates among those who attempt to resolve conflicts between groups and nations (Barry & Oliver, 1996; Coleman, 2012). The first order of business is diffusing the immediate tension and trying to create a dispassionate state of mind in the parties to the conflict. To be sure, people in the grip of an intense emotional state are unlikely to seek compromise or adopt one another's perspective. But this does not mean that the emotional underpinnings of the conflict should be minimized in importance or marginalized when trying to understand the source of the conflict or when trying to predict or control the outcome of the conflict scenario. Emotions are the fuel and lifeblood of intense conflicts, particularly of those that seem intractable, so not taking them into account is likely to provide a very limited and misleading portrait of the conflict.

People have a wide repertoire of emotions and each one signals something different about the underlying concerns and how they can be effectively addressed. Basic emotions such as anger and fear are often in reaction to material threats and uncertainty, for example, whereas the *self-conscious emotions*—embarrassment, humiliation, guilt, shame, and pride—are reactions to threats to one's personal or collective identity (e.g., personal worth, the status of one's ingroup). Thus, knowing the salient emotions experienced by people in conflict may provide important information about the source of the conflict, the factors that maintain it, and possible avenues of resolution. If humiliation is the dominant emotion for one of the parties, for example, focusing solely on issues of scarce resources—which may have initiated the antagonism—may miss the point entirely about why the conflict is sustained and seemingly intractable (for a research example see: Coleman, Goldman, & Kugler, 2009). One's efforts instead should focus on reducing feelings of humiliation.

In the dynamical perspective, emotions have a reciprocal causal relation with attractor dynamics. Strong emotions, first of all, intensify people's natural press for coherence and thus promote the emergence of deep and rigid attractors for destructive conflict (Vallacher, Coleman, Nowak, & Bui-Wrzosinska, 2010). But once an attractor is established, it can unleash strong emotions when it is activated by conditions, and especially when it is threatened by information that seems to call into question its validity. Beyond that, the nature of emotion can shape the content of attractors. Anger, for example, can promote a single-minded view of interpersonal or inter-group behavior reflecting revenge or physical harm. An attractor that emerges from perceived humiliation, meanwhile, might reflect a single-minded focus on repairing one's status in the eyes of others or denigrating the personal qualities and status of the source of humiliation. In analyzing conflicts, then, the nature and intensity of emotion should be a major focus of attention rather than marginalized in importance.

Overly Simplistic Modeling

Parsimony is a central value in every area of science. In constructing a theory, one should employ as few assumptions as possible and frame the phenomenon at issue in terms of a minimal number of principles and corollaries. And when two competing theories are advanced for a phenomenon, parsimony dictates that we favor the theory that "explains the most with the least." As Kuhn (1970) and others have observed, science does not necessarily advance by discrediting old theories—although that certainly happens—but rather by the acceptance of a new theory that provides a more parsimonious account of how and why things happen.

It principle, then, the development of simple models of conflict is a reasonable goal. In practice, however, a concern with parsimony can lead scholars to ignore many important considerations that are relevant to the emergence of conflict and that might be important in conflict resolution. Apart from the risk of ignoring the web of factors and conditions associated with conflict, adherence to parsimony poses problems because of the way it is practiced. For many, parsimony is an invitation to reductionism—the assumption that processes at a given level are isomorphic with the processes at a lower level. In explaining antisocial behavior, for example, one could invoke the same principle (e.g., frustration-induced aggression) at both the interpersonal level (i.e., between individuals) and the societal level (i.e., between groups). As noted in the discussion of dynamical minimalism (Chap. 3), however, reductionism is often an unwarranted assumption. It may be parsimonious to see the same forces at work at different levels, but this perspective does not pay homage to the potential for emergence in complex systems. The processes of mind are certainly rooted in brain dynamics, for example, but the nature of mental processes—hopes, fears, inductive logic, etc.—cannot be found in neurons and synapses.

Overly Complex Modeling

Many peace and conflict scholars have recognized the dangers of over-simplification in attempting to understand the complex and diverse nature of conflict. Highly sensitive to the intricate web of forces at work, theorists often incorporate as many variables as possible in their models, so as not to miss something that may prove critical to deep understanding. General systems theory (Von Bertalannfy, 1968), for example, was embraced as a way of incorporating literally dozens of factors and mapping their connections to obtain a holistic model of social relations generally and conflict in particular (Blalock, 1989; Fisher & Keashly, 1991; Sandole, 1999). Such an approach may produce a very comprehensive description of conflict, but it hardly satisfies the requirement of parsimony, nor does it provide a deep level of insight into the basic underlying forces that surface in conflicts that differ with respect to relevant factors and conditions.

This state of affairs sounds like a paradox. We shouldn't rely on simplistic theories because they overlook important factors, but neither should we rely on complex theories because they lack parsimony and do not expose the fundamental forces at work. Fortunately, the approach of dynamical systems provides a way out of this dilemma. Common to dynamical systems in all areas of science is the potential for emergence. This means that complexity at a macro level does not necessarily require complexity at a lower level. To the contrary, often a small number of very simple rules at a basic level of system functioning can promote the emergence of very complex properties and processes at higher levels of system functioning. This is the guiding principle of the afore-mentioned approach of dynamical minimalism. In essence, this approach to theory construction and testing enables students of conflict to embrace both the rule of parsimony and the reality of complexity in real-world conflicts.

Missing the Invisible

The manifest nature of conflict is hard to ignore. When individuals or groups engage in antagonistic behavior toward one another, it is quite understandable that theorists and practitioners would attend to that behavior and the obvious triggers in the immediate context. When someone is insulted and responds aggressively to the insulting party, for example, one is likely to see the conflict primarily in these terms. The focus on the visible features of a conflict is obviously important, but this can blind one to the more subtle undercurrents and longer-term trends that characterize the relationship between the parties. Indeed, the visible features may provide a misleading portrait of the real dynamic processes at work. The person who responds aggressively to a perceived insult, for example, may really be acting on the basis of very different factors—perceived injustice, perhaps, or ideological differences—with the insult simply providing a tipping point or excuse for hostile behavior.

Focusing solely on the visible features of a conflict is problematic for another reason. In a dynamical system, there may be more than one attractor available for capturing the system's dynamics. Only one attractor is manifest at a time, however, in the same way that a reversible figure-ground relationship (e.g., a Necker cube) is seen from only one perspective at a time. The other attractor is latent, but under the right conditions it can become the dominant mode of thinking and action. So although latent attractors are invisible, they nonetheless represent potential coherent states to which the system can converge. To truly understand the nature of a relationship marked by conflict, then, it is critical to identify the invisible (latent) states as well as the visible (manifest) states of the relationship system. Knowing the attractor landscape in its entirety, moreover, provides insights into how to transform a destructive relationship into one that is benign or even constructive.

Overlooking Unintended Consequences of Well-Intentioned Acts

One of the cruel ironies of life is that well-intended actions sometimes produce bad outcomes. Forgiveness in a marital relationship, for example, may be motivated by empathy and a desire to keep the relationship intact and it generally has the intended effect. However, if the transgression prompting forgiveness was particularly egregious and the transgressor has acted in similar fashion on previous occasions, the act of forgiveness can reinforce a pattern of behavior that is likely to rupture the relationship rather than heal it (e.g., McNulty, O'Mara, & Karney, 2008). There is evidence that the unintended and ironic consequences of well-intended behavior play a surprisingly large role in conflict scenarios. In fact, according to Dietrich Dorner, a psychologist who studies leadership and decision making in complex environments, more harm is done in today's world by people who are trying to do good but who are unaware of the long-term consequences of their behavior than by people who knowingly act in a destructive or malicious fashion (Dorner, 1996). In trying to solve immediate problems, for example, negotiators and mediators are likely to neglect the more pressing but long-term issues, which can fester and become magnified with the passage of time.

This scenario is easy to appreciate from a dynamical perspective. A destructive relationship evolves in accordance with various temporal patterns, making it difficult to anticipate the manner in which an act early in the sequence will be manifest at a later time. This is particularly likely if the relationship is governed by a strong attractor that can transform the meaning of any action to make it fit the theme of the attractor. An act of generosity toward an opponent may be accepted at face value initially, but over time it might be seen by the opponent with a deeply entrenched attractor for negative judgments as a devious attempt at manipulation. So although the act was well-intended, over time it reinforces the opponent's hostility and can prompt negative behavior rather than reciprocity or gratitude.

Ignoring Relevant Theory and Research

Conflict, especially conflict that progresses to intractability, certainly seems like a unique aspect of human relations. People spend the vast majority of their time getting along reasonably well with others, and when they do clash the conflict is usually short-lived. So perhaps conflict is a special topic that requires a correspondingly unique explanation. This, however, is not the lesson of either classic or contemporary social psychology. Since the field's inception, theorists and researchers have explored the fundamental of conflict and have found them in basic models of mind and action. Among the topics with direct relevance to difficult conflicts are ingroup–outgroup relations, the formation and maintenance of stereotypical thinking, the assortment of cognitive biases that shape and maintain prejudice, the assumptions at work in the perception of other people's minds, and the role of self-esteem, social identity, and ideology in social relations. These mundane processes of mind and action need to be incorporated into the heart of theories and research programs concerned with conflict.

As should be readily apparent by now, the scholarship relevant to the dynamics of conflict extends far beyond social psychology. Indeed, the fundamental principles at work in the emergence, maintenance, and resolution of conflict can be observed in systems throughout nature. Intractable conflict may be infrequent in human relations, but a deep, comprehensive, and heuristic account of this phenomenon requires an understanding of basic processes that unite phenomena of all kind. To be sure, the translation of the principles of nonlinear dynamical systems into concrete means of resolving enduring rivalries and other expressions of intractable conflict is hardly a trivial matter. The basic elements of this translation are the focus of the next section.

Changing the Attractor Landscape

In daily life, we occasionally encounter someone with an attitude that appears maladaptive or simply wrong, or who acts in a manner that is harmful to others and only serves to reinforce destructive interpersonal relationships. The temptation in such instances is to point out the fallacy of the attitude or the negative consequences of the behavior. Yet we know that such confrontations are rarely effective. More often than not, this strategy of changing minds and actions either falls on deaf ears or produces a defensive reaction and thus proves to be counterproductive. In the same vein, attempts to challenge directly the validity or instrumental value of a strong and wide attractor for hostile relations are doomed to fail and are likely to intensify the beliefs and response tendencies of the person or group whose dynamics converge on the attractor. Inherent in the definition of an attractor, after all, is resistance to perturbing influences and enhanced efforts at maintaining the pattern of thought and behavior that is threatened.

From the perspective of dynamical systems, bringing about meaningful and lasting changes in a relationship mired in destructive conflict requires changing the attractor landscape constraining the thoughts, feelings, and actions of the parties to the relationship. At best, a direct assault on the attractor will promote a change that is likely to be short-lived and ultimately resisted. A change in the attractor landscape, however, essentially reconfigures the ensemble of likely states that can be adopted and alters the forces operating within the system. Changing the attractor landscape, then, can bring about changes in system behavior that are not only lasting but also self-sustaining over time.

In principle, there are three basic scenarios by which the dynamics of an intractable conflict might be transformed and create the potential for benign or even positive interpersonal and intergroup relations: *disassembly and reconfiguration of an existing attractor*; *activation or creation of a latent attractor*; *change in the number and types of attractors*. These scenarios are discussed in turn below.

Attractor Deconstruction

This scenario captures the Lewinian insight of "unfreeze—move—refreeze" discussed in Chap. 3. Attractors represent states of mind and action orientations that are "frozen," unyielding in the face of new information and events. Yet, they mask underlying dynamics that are highly volatile. By tapping into this volatility, one may be able to unfreeze the system, enabling it to explore other higher-order integrative states. When a desired higher-order state is expressed during the "move" phase, efforts can be directed at locking it in, effectively "refreezing" the system at this new attractor.

To see how this general strategy can be translated into practice, it is useful to recall how attractors are formed—and then reverse the process. Attractors emerge from the nonlinear interactions of reinforcing and inhibitory feedback, so that the lower level elements comprising a system influence one another to take on the same global meaning (e.g., evaluation). In practice, this means that separate issues, events, and pieces of information become linked by reinforcing feedback to promote a coherent (internally consistent) perspective and an unequivocal action orientation. In losing their independence and interpretative elasticity, the elements are protected against change in meaning, analogous to the way members of a group come to the defense of a group member who is threatened by someone from outside the group. To "reverse engineer" this process means deconstructing the attractor by changing some of the feedback loops from reinforcing to inhibitory or by restoring independence among the elements.

One strategy is to isolate individual elements so as to de-emphasize their integration with other elements in the system. Research conducted within the framework of action identification theory (Vallacher & Wegner, 2012) illustrates this "disassembly process" in the personal understanding and control of behavior. Individuals who have a high-level identity for what they are doing tend to resist alternative interpretations of their behavior. But if they are induced to think about their action in terms of a small set of lower-level identities (e.g., specific motor movements or utterances), their understanding of what they are doing (or have done) becomes fragmented and open to reconfiguration. This is the essence of the emergence process that is central to the theory.

There are different means of inducing lower levels of action identification (cf. Vallacher & Wegner, 1987). Anything that disrupts an ongoing action, for example, makes a person sensitive to the over-learned details of his or her action at the expense of the action's higher-level goal or meaning. From this lower-level state of mind, the person becomes open to new ways of coordinating the lower-level features of the action, so that the action can be understood and controlled with a new goal, plan, or sensitivity to consequences. People who are asked to drink coffee from a heavy, unwieldy cup, for instance, become sensitive to the mechanics of drinking (e.g., holding the cup, bringing the cup to one's mouth). Even among experienced coffee drinkers who normally monitor this action with respect to a personal higher-level meaning (e.g., becoming alert, starting the day), this lower-level focus promoted susceptibility to subsequent messages that suggested different consequences and implications of drinking coffee. Depending on the message they received, they were led to believe that coffee drinking makes one seek out further stimulation or that it has precisely the opposite effect—that it makes one avoid further stimulation (Wegner, Vallacher, Macomber, Wood, & Arps, 1984). This openness to social influence was not observed for participants who drank from normal cups, presumably because they maintained a focus on their personal higher-level meaning for this act.

Deconstruction of a higher-level meaning also occurs when people are instructed to focus on the details of a narrative rather than on the narrative's larger meaning (e.g., its theme or implications). Focusing on the specifics renders the person vulnerable to communication from others regarding the larger meaning of the narrative. Vallacher and Selz (1991), for example, asked participants (both males and females) to read a lengthy account of an alleged rape incident. Compared to participants who focused on the overall meaning of the narrative, those who were instructed to focus on the specific acts in the narrative subsequently embraced whatever conclusion was provided by the arresting offer regarding the allocation of blame for what happened. They blamed the alleged rapist for the incident if the officer concluded there were sufficient reasons to press charges against him, but they tended to blame the victim if the officer concluded that there were insufficient grounds for doing so. In short, when habitual actions and generalized beliefs are deconstructed by focusing on lower-level features of an action or an event, people become vulnerable to new interpretations that provide an avenue of emergence to a new coherent perspective.

This deconstruction-reconfiguration scenario corresponds to the disassembly-emergence process of dynamical systems, described in Chap. 3, and thus may extend beyond the personal understanding and control of action to describe the dynamics of other psychological and social systems. With respect to inter-group conflict, this strategy could be implemented by focusing each party's attention on specific actions, events, and pieces of information, but doing so without noting their connection to the larger pattern in which they are embedded. When decoupled in this way, the lower-level elements may become re-configured into an entirely different pattern (e.g., a positive or sympathetic view of the outgroup). This strategy might require singling out individuals in a group in order to block the combined influence of other group members that would reinforce the individuals' preconceived beliefs and feelings. With the reinforcing feedback from other group members cut off, the disassembly-reconfiguration scenario could be implemented to produce a new way of thinking and acting with respect to the outgroup.

Yet other strategies playing on the emergence process can be envisioned and are worthy of consideration. The key lesson here is straightforward. Attacking a pattern of belief or action is likely to intensify rather than weaken the pattern because of the tendency for attractors to resist perturbing influences. Instead, the strategy should be isolation of the system's elements, thereby weakening or eliminating the reinforcing feedback loops among them, and providing an avenue of emergence to a new way of thinking and acting that provides coherence for the elements.

Creation of Latent Attractors

The second scenario capitalizes on the tendency for latent attractors to develop in dynamical systems. The basic idea is to move the system out of its manifest attractor of hostile sentiment and interaction patterns into a latent attractor that is defined in terms of benign or positive thoughts, actions, and relationships. This process actually parallels a basic principle of Gestalt psychology. Thus, in much the same way that perceptual elements can be organized to promote different holistic patterns, social information and events can take on a host of different meanings with diverse action possibilities. There are likely to be many elements of relevant information (facts, events, rumors, memories, etc.) in an interpersonal or inter-group relationship, and these elements can be configured in different ways to achieve a coherent perspective and a stable platform for action. At any one time, though, only one of the potential attractors is likely to be manifest in people's orientation toward other people or groups—just as a single figure-ground relation is perceptually salient at a given point in time. The other potential attractors are latent and may be effectively invisible to observers and perhaps to the parties involved as well. Under some circumstances, however, these latent attractors may become manifest and promote a corresponding change in the relations among people and groups.

Ironically, latent attractors may actually be common in relationships whose manifest dynamic tendencies are frozen in a global state reflecting hostility. This is because protracted conflicts often involve a long history between the parties, including stretches of time characterized by positive relations. The relations between Ukraine and Poland are a case in point. The relations between these nations go back many decades, and during this time there were periods marked by mutually beneficial trade and cultural relations. In fact, the strained relations that developed as a result of World War II could be viewed as the exception rather than rule in Ukrainian-Polish relations. So despite the manifest antagonism between these two countries over the past half century, there is a strong latent attractor for positive relations. From this perspective, it is understandable that the mutual antagonism gave way in a dramatic fashion to mutual support during the Orange Revolution in Ukraine in 2004–2005. In effect, the two countries experienced a shift in their respective figure-ground relations and began to interact in a qualitatively different manner that reflected an attractor that had been latent for years.

Capitalizing on latent attractors may even be a viable strategy when attempting to resolve conflicts among stakeholders with a history of largely negative relations. If a latent attractor for positive relations does not exist, there may be ways of creating one. A hopeful example is provided by Staub's (2011) recent peace building and reconciliation work in Rwanda and the Congo, which illustrates how latent positive attractors can be constructed through public media campaigns. The goal of these campaigns is to establish a positive shared reality and to educate people on the forces responsible for the emergence of hostilities. The situation in these countries may not be ripe for change at this point in time, but changes in political conditions may allow for a latent attractor of peace created in this manner to become manifest.

Even in the absence of a long history of relations, the development of a latent attractor is conceivable. Indeed, the emergence of latent attractors may be an inevitable consequence of developing a manifest attractor. In order to maintain a coherent orientation toward another person or group, people necessarily suppress or discount inconsistent ideas, feelings, and action tendencies. A peaceful overture by a disliked outgroup, for example, can prove difficult to reconcile with the ingroup's rigid view of the outgroup and may be discounted as irrelevant. Once new information enters a system, however, it tends to persist in some fashion, even if it is inconsistent or unwanted (cf. Wegner, 1994). Suppressed facts and ideas may exist at the fringe of consciousness, but over time they may undergo self-organization and set the stage for the emergence of a new attractor for thoughts and feelings. This idea is consistent with recent theory and research on nonconscious thinking (e.g., Bargh, 1996; Dijksterhuis & Nordgren, 2006; Murphy & Zajonc, 1993; Wilson, 2002) and implicit attitudes (cf. Greenwald & Banaji, 1995). So despite the attempts to discount the outgroup's peaceful overtures, these inconsistent elements may eventually coalesce into a new attractor reflecting positive attitudes toward the outgroup. At this point, the system may be ripe for a change in its manifest dynamics. If an intervention temporarily defuses the conflict, for example, the newly formed latent attractor could become manifest and reshape the pattern of thoughts, feelings, and actions in the relationship.

The existence of latent attractors illustrates the potential for nonlinear—rapid and qualitative—change in the state of a system that appears to be rigid and impervious to influence.

Intractable hostility between two groups can be suddenly transformed to relatively benign or even positive relations if an event pushes the relationship out of its current basin of attraction into the basin of a previously latent attractor. This suggests the need to take a long view when attempting to resolve difficult conflicts. Peacekeeping missions, reconciliation processes, and trust-building activities often appear to be ineffective in situations characterized by a protracted struggle between groups. But these conflict resolution initiatives may have the effect of establishing or bolstering a wide and deep attractor basin for moral, humane forms of intergroup interactions, which can provide the foundation for a stable, peaceful future when the situation is ripe for such transformation. The creation and activation of latent attractors may be imperceptible, but this process can create the potential for a positive state that would be impossible otherwise.

Keep in mind, though, that latent attractors represent a double-edged sword. In the same way that negative relations between groups can undergo a dramatic transformation to positive relations, the opposite scenario is just as plausible in principle. If conditions support a latent attractor comprised of negative thoughts, feelings, and action tendencies, manifestly peaceful relations between groups could suddenly change to a hostile pattern of intergroup behavior.

Changing the Number and Type of Attractors

The first two scenarios are implemented within an existing attractor landscape. It is also possible, however, to transform the landscape itself and in effect change the potential states of the system. This is the basis for the third scenario: changing the number and type of attractors in an interpersonal or intergroup relationship. Many factors can promote quantitative changes in a system's behavior (e.g., moving the system from a manifest attractor to a latent attractor). However, only a small subset of these factors—referred to as *control parameters*—can promote *qualitative changes* in the system's behavior (see Nowak & Lewenstein, 1994; Ruelle, 1989). Changes in the attractor landscape are called *bifurcations*. Bifurcations can be manifest in several different ways: a change from a single attractor to two attractors, a change from a single attractor to a periodic attractor (oscillation between two or more coherent states on some timescale), and a sequence of changes from a single attractor through periodic and multi-periodic attractors to a chaotic attractor (a complex trajectory of behavior that never repeats and is highly sensitive to initial conditions). For each type, change occurs when the value of the relevant control parameter reaches a specific threshold.

The bifurcation scenario has considerable potential for changing interpersonal and intergroup dynamics in a fundamental way. But translating this general idea into a specific strategy for conflict resolution can prove difficult. First, one must identify the system's control parameters out of the ensemble of factors that are relevant to the conflict in question. Is enhanced law enforcement, for example, something that simply changes the current state of a conflict between mutually antagonistic groups? Or does law enforcement have the effect of changing the number and type of potential states for inter-group thoughts, feelings, and action?

Once the system's control parameters are identified, the task is determining the form that bifurcations are likely to take when the values of the control parameters are changed. Let's assume law enforcement holds potential for transforming the attractor landscape of an inter-group conflict. What type of bifurcation can be expected when in response to changes in police security? Perhaps a notable enhancement in police security maintains the existing attractor (hostile inter-group relations) but also adds a second attractor (e.g., hostile relations with the local government). This type of change was observed in 2005, when the French government sent swat teams into immigrant neighborhoods to quell unrest reflecting tensions with the non-immigrant population. Alternatively, enhanced security could promote the emergence of a new attractor positive relations between groups, with both groups recognizing a common enemy. Yet another possibility is that changes in police security could produce periodic movement between different attractors (e.g., oscillation between positive and negative relations) or even chaotic movement over time between very different patterns (e.g., seemingly random swings in relations that are highly sensitive to minor events).

Clearly, identifying a system's relevant control parameters and specifying their effects on the attractor landscape in a conflict scenario represent challenging theoretical and practical tasks. Such efforts may have a significant payoff, however. Indeed, transforming the attractor landscape of a relationship mired in conflict may prove to have a longer-lasting effect on a relationship mired in conflict than would disassembling a manifest attractor or activating a latent attractor that is already part of the attractor landscape.

Implications for Conflict Management

Changing the attractor landscape of an enduring and destructive conflict constitutes a radical departure from the standard practices of conflict management. This departure comes into sharp relief when the three scenarios we have described are unpacked into a set of specific guidelines. Below are ten such guidelines that follow from our account (adapted from Coleman, 2011).

1. Analyze and respond to the unfolding dynamics of conflict, not merely to discrete actions or events.
2. Envision elements of conflict systems (parties, issues, events, etc.) as mutually interconnected and evolving through reinforcing and inhibiting feedback loops rather than focusing on casual arrows or links.
3. Aim to alter the longer-term patterns of dynamics between stakeholders in a conflict system rather than seeking only to achieve outcomes such as treaties and agreements.

4. Privilege the importance of the emotional context of conflicts. Recognize the "reservoir" of positivity and negativity that builds up over time in relationships and the critical role of this reservoir in shaping the interests, thoughts, and actions of disputing parties.
5. Think differently. Encourage and attempt to integrate contradictory bits of information into the understanding of events and decision-making processes. Be vigilant regarding the seductive press for coherence and tendency toward over-simplification in polarized conflicts.
6. Remain mindful that conflict and peace often coexist. They do not exist on a single bipolar dimension, but rather operate in ways that are somewhat independent of one another. Conflict and peace represent two alternative realities, with one manifest and the other latent at given points in time. The presence of war does not mean the absence of peace—and vice versa.
7. Attend to latent (invisible) potential. Pay attention to anomalies and surprises that occur, for they may signal cracks in the foundation of the current understanding of a conflict and suggest that alternative attractors are operating beneath the radar.
8. Open up the system. Foster understanding and relations that are characterized by higher levels of emotional, cognitive, behavioral, and structural complexity. This increases the capacity for movement and constructive change.
9. Develop strategies based on opening up the system. Identify actionable solutions that are informed by awareness of the higher levels of complexity.
10. Anticipate and be prepared for unintended consequences. Adaptation is a continual process that requires remaining open to feedback, with willingness to rethink key decisions and alter courses of action. Such flexibility is crucial in a complex and unfolding relational context.

Practical Implications for Addressing Seemingly Intractable Conflict

The theory and research described in this book offer a wealth of ideas for disputants, other conflict stakeholders, analysts, negotiators, mediators, diplomats, and other third-parties to help re-orient their understanding of intractable conflict. Ultimately, however, the value of these ideas will be measured by their ability to be translated into practical strategies and skills for changing the dynamics of destructive conflict. In several of our previous publications (Coleman, 2011; Coleman, Bui-Wrzosinska, Vallacher, & Nowak, 2006; Coleman, Fisher-Yoshida, Stover, Hacking, & Bartoli, 2008; Coleman, Vallacher, Bartoli, Nowak, & Bui-Wrzosinska, 2011; Nowak et al., 2010; Praszkier, Nowak, & Coleman, 2010), we have begun to give shape to a new methodology for action; a set of sound, evidence-based strategies and tactics for addressing enduring conflicts. These are organized around a set of three basic conflict management practices (Coleman, 2011), which are aimed at constructively managing the current state of the conflict while increasing the probabilities for constructive relations between the parties in the future *and* decreasing the probabilities for destructive future encounters. They practices include:

1. **Complicate to simplify**. This practice entails breaking through the press for coherence and oversimplification that is so forceful and constraining with polarized disputes, while attempting to identify the actionable hubs, gateways, and patterns otherwise disguised by the complexity of the conflict. This constitutes a framework for *conflict analysis and feedback loop mapping* which can allow for new understandings and insights to emerge for addressing difficult conflicts. In sum, complexity matters.

   Learning to effectively map the dynamics of intractable conflicts is one approach to capturing their complexity. It typically requires us to both see these problems in new ways *and* to attend to aspects of the problems we are not used to seeing. Of course, the high degree of complexity of these situations often defies comprehensive analysis. But that is not the objective of this practice. Feedback loop–mapping can allow contextualization, discovery and insight to emerge; our sense of the issues and events in a conflict can take on new meaning when seen from such a "field" or relational perspective (see Lewin, 1936).

   However, at the heart of this practice is the double-edged sword of complexity. If the narrative of a conflict has lost complexity and become overly simplistic and polarized, we must reintroduce nuance into the understanding of the situation. But adding too much complexity can easily make an already complicated situation seem overwhelming and immobilizing. Therefore, in addition to mapping complexity, it is critical to employ a few general tactics for *managing complexity*. These are informed by the approach of *dynamical minimalism*, the aim of which is to make use of the complexity of a problem in order to find the minimal set of mechanisms that can account for its evolving pattern, which represents its essence. In other words, the goal is *simplicity informed by complexity*.

Capitalizing on Complexity

Here are five actions for harnessing the complexity generated by conflict-mapping, which can help to identify more focused strategies:

1. **Identify hubs, loops, and energy in the system**: Beginning by identifying central *hubs* of activity, elements that link with many other elements; key *reinforcing loops*, elements that stimulate themselves through links with other elements in an ongoing fashion; and the *ratio*, or balance, between conflict-reinforcing and -inhibiting feedback, can help determine whether a conflict is escalating, deescalating, or stuck in a stalemate. Identifying central hubs and strong feedback loops is important for locating *centers of energy in the system*, *gateways*

for high-impact intervention, *strategic targets for introducing conflict-inhibiting feedback* (such as early-warning systems that deter escalation), and *peace-reinforcing feedback* (like high-stakes common interests that motivate reconciliation).

2. **Identify local actionables**: It can also be immensely helpful to focus in on more local elements and links that are *actionable*; that can feasibly be addressed. Feasible actions, informed by the complexity of events, can affect the probabilities of constructive change in a conflict system, but they should only be targeted after a fuller mapping of the system has provided a sense of the context in which these elements are operating.
3. **Locate what is already working in the system**: There are usually processes and mechanisms already operating in every conflict system, no matter how entrenched, that are functional and constructive. They are often be overlooked in conflict analyses, as the suffering and crises make the *problems* salient, usually at the expense of existing buffers or remedies. But a thorough scan of the system can help to locate its more functional components; including local norms and practices that prohibit aggression and violence beyond certain levels, indigenous grievance systems or other regulatory mechanisms considered impartial and fair, or widely respected members of families or communities that might be able to play a more actively constructive role in addressing disputes.
4. **Identify integrative catalysts**: These are individuals who embody the different conflicting identities and tensions inherent to the conflict. They may simply be trusted friends or family members, or have multicultural or multiethnic identities, but typically they hold a special status that allows them to stay above the fray and not get forced into one side or the other—and therefore can play a critical facilitating role.
5. **Visualize the attractor landscape** (see below): Conflict maps provide a sketch of the *process architecture* of a conflict: like taking an X-ray of the internal workings of the body, an excellent tool for diagnosis and treatment planning. Ultimately, however, we have found it useful to move from mapping to working with a simple visualization software program to begin to see how the different elements of a conflict interact together *over time*. This is critical for focusing our understanding on how the conflict system evolves and establishes temporal patterns or attractors across time.

2. **Build up and tear down**. This is the practice of stirring or creating a sense of hope and possibility for more constructive relations in a conflict, while simultaneously deconstructing the traps for enmity and destruction that lie in wait to recapture the dynamics of the system. These tactics constitute the main category

of *strategic action* for the DST approach (*disassembly and reconfiguration of an existing attractor*; *activation or creation of a latent attractor*; *change in the number and types of attractors*); and involve an array of tactics for altering the conflict attractor landscape over time.

Here the idea of *latent attractors* comes into play. We have found that the malignant thoughts, feelings, and actions that characterize a person's or group's dynamics in intractable conflict represent only the most obvious attractor for the conflict. In particular, where there is a long history of interaction between people or groups, other potential patterns of mental, emotional, and behavioral engagement typically exist, including those promoting positive relations. But latent attractor dynamics are hard to see. Some changes in a system can be easily observed because they affect the current (overt) state of the system. Other changes, however, may only affect the *possible* states of the system and thus not be immediately apparent. Such changes may remain latent for extended periods of time, yet manifest rapidly in response to external influences and events that seem relatively minor.

Although movement between attractors may be rapid and abrupt (back and forth between constructive and destructive patterns), *the change of attractors themselves is likely to be far slower and more gradual*. Thus, when a specific policy or intervention does not produce a visible effect, this does not mean that it is futile. Rather, such activities may be creating, deepening, or destroying latent attractors in a system. In other words, they may affect the range of possible states rather than the current state.

With this in mind, identifying and reinforcing latent positive attractors—"traps" for peaceful or constructive relations—should be among the principle aims of conflict prevention, intervention and repair. The identification, support, and initiation of constructive forces within the system (common projects, citizen exchanges, dialogues, etc.) are critical for increasing the long-term chances that peaceful relations will resurface—*whether they show short-term results or not.*

Building Potential

A few of the many actions available for creating and bolstering positive latent attractors include:

1. **Circumventing conflict**: Typically the destructiveness of protracted conflict exacerbates the very negativity and strife that created the conflict conditions in the first place, and thereby perpetuates it. However, attempts to address these circumstances directly, in the context of a peace process, typically elicit resistance; they are seen as affecting the balance of power in the conflict (usually by supporting lower-power groups most affected by the conditions). Interveners recognizing this will work to address these conditions of hardship, without making any connection whatsoever to the conflict or peace processes. This

unconflict resolution strategy can help address some of the negativity and misery associated with conflicts, without becoming incorporated (attracted) into the polarized "good versus evil" narrative of the conflict (see Praszkier et al., 2010).

2. **Employ weak power**: Sometimes peace does emerge out of long-term conflicts, and we suggest that one reason may be the power of powerlessness. That is, the unique influence people and groups with little formal or "hard" power (physical strength, military might, finances) but effective "soft" power (trustworthiness, moral authority, wisdom, kindness) can have in these settings. Weak-power third parties are at times able to carefully introduce a sense of hope for change in the status quo, a sense of doubt or dissonance in an ultra-coherent "us versus them" meaning system and can begin to model other more constructive means of conflict engagement (shuttle diplomacy, mediation, negotiations).
3. **Construct chains**: A tactic designed to break through deadlocks and initiate peace talks in protracted conflicts is the use of negotiation chains; engaging a sequence of actors in the exploration of more formal talks. Actor A speaks directly to actor B; A is not politically constrained against speaking to B, but B has contacts further down the chain with the other side. This allows talks to transpire through a series of encounters between parties who (1) need to be able to maintain deniability in the talks, and (2) would otherwise not be able to communicate.
4. **Identify superordinate goals and identities**: Seen from the perspective of attractor landscapes, finding common ground between parties, emphasizing shared goals and concerns, facilitating trust-building activities, and incentivizing cooperative conflict resolution initiatives—although they may appear to be largely ineffective in situations locked in an ongoing protracted struggle—may in fact be acting slowly and indirectly to establish a sufficiently wide and deep attractor basin for moral, humane forms of intergroup relations. One day, these initiatives may provide the foundation for a stable, peaceful future.
5. **Leverage the irony in impossible conflicts**: Very strong conflict attractors will often exclude a wide variety of information discrepant from the main point of view. An explicitly peaceful overture by an out-group, for example, is difficult to reconcile with the in-group's negative attitude, and thus may be discounted as an anomaly. Should enough incidents like this occur, however, they may begin to coalesce into a new or latent attractor reflecting benign or positive attitudes toward the out-group. At this point, if an event or intervention temporarily defuses the conflict, the newly formed latent attractor could suddenly become

(continued)

manifest and redirect the in-group's thoughts, feelings, and actions vis à vis the out-group.

The opposite of building up positive attractors is working actively to break down destructive conflict attractors, which are usually still operating in a latent manner even after more constructive relations have re-emerged. The objective here is to make negative attractors *less attractive*. Even when conflict intensity diminishes and disputants appear to move into a state of relative calm, it is critical that we recognize that the lure of destructive interactions still exists. Attractors often remain as potent and alluring as ever when latent, even at times increasing in strength. It is critical, then, that we recognize when coming out of a conflict that the current state of peace is most likely temporary and therefore work actively to deconstruct and dismantle the infrastructure of negative attractors.

Breaking Down Hidden Traps

There are a variety of actions useful for dismantling negative attractors, including:

1. **Work down below:** Research has shown that global mental states such as strong identities or attitudes, or even fervent beliefs regarding "truths", can be effectively disassembled into their lower-level elements; that is, into the different component parts that constitute the attitude or belief. Doing so can create the potential for a wholesale change in people's understanding of their own and others' actions. With conflict, this means calling attention to specific actions, events, and pieces of information without making the connection to the global pattern in which they are embedded. When decoupled in this fashion, these lower-level elements can become reconfigured into an entirely different pattern (e.g., a more positive view of the out-group and benign interaction pattern). It is like making small adjustments in the mechanics of a tennis stroke or a golf swing that lead to radical changes in the player's overall game.
2. **Institutionalize more nuanced conflict narratives**: Intractable conflicts typically result in (and are perpetuated by) biased narratives in families, schools, communities, and the media about the history of the conflict; who played the roles of hero and villain and what is still at stake. Establishing mechanisms to monitor and revise such one-sided narratives are essential for preventing future generations from returning to the same destructive patterns.
3. **Limit the accumulation of negativity through movement**: Negativity is a potent force, possessing the impact of about five times that of

positive experiences. Research on trauma has shown that negative encounters have a tendency to accumulate, both psychologically and socially, over time, particularly when groups are ghettoized. Our research conducted through computer-based modeling has led to the paradoxical finding that *movement* is key to dissipating negativity in community relations. Systems where negativity is relatively unconstrained, and where members of groups are allowed to travel and disperse, will tend to show a dissipation of negativity over time.

4. **Foster repellers for violence:** Most communities have laws and other prohibitions against physical violence, establishing what are known as *repellers* in complexity parlance. In fact, archeological research suggests that communal taboos against violence have existed for the bulk of human existence, and were a central feature of the prehistoric nomadic hunter-gatherer bands. Indeed, a key characteristic of peaceful groups and societies, both historically and today, is the presence of nonviolent values, norms, ideologies, and practices.

3. **Change to stabilize.** This practice views adaptation and change as critical to stability and sustainability of peaceful relations. It is about leveraging opportunities for change in a manner that is mindful of non-linearity, long-term dynamics and the need for adaptation in complex systems, and highlights the importance of adapting effectively to changing circumstances in order to achieve sustainable solutions. It concerns understanding the things we need to do in order to produce a *qualitative change* in the current state of the conflict (moving the conflict between attractors) or in the character of the landscape. It is also about knowing when social conflicts are *ripe* for change; discovering those *few things that change everything* (altering attractor landscapes), and about *understanding the difference*, in terms of time and impact, of various change strategies. But it is mostly about fostering stability through sustainable change, which comes from enhancing our abilities to adapt.

Stabilizing Change

A few specific tactics useful for managing change/stability include:

1. **Leverage Instability**. Coherence and stability in social conflicts are not the enemy until absolute certainty about "us versus them" takes over and provides *the* foundation for understanding; then it's probably time to change things up. One leading way that instability occurs in these conflicts is through major shocks to the system (world wars, civil wars, regime change, an independence movement, etc.). Over three-quarters of the enduring conflicts that ended between 1816 and 2006 did so

(continued)

within ten years of a major political shock (Goertz & Diehl, 2001, 2006). Such shocks destabilize conflict systems and allow for the deconstruction and reconstruction of the attractor landscape, although this may take years to become evident, as the initial shock most likely affects factors which affect other factors and so on until overt changes occur.

2. **Grow Ripeness**: *Ripeness*, defined as a willingness of disputants to negotiate peace, is not only determined by the perceptions and motives of the parties but also by the often vast constellation of forces that make up the attractor in which they are trapped. Conflict is only truly unripe for resolution if the attractor representing destructive conflict is very potent and no alternative attractors exist. A conflict begins to ripen for resolution when an alternative attractor is fostered or develops, corresponding to some form of peaceful engagement that may eventually capture the dynamics of the system. The conditions and processes that can *increase the probabilities of ripeness* in long-term conflicts include (1) the many psychological, social and structural factors that strengthen positive attractors for thinking, feeling, and acting between the parties, (2) the many factors that can weaken negative attractors, and (3) the factors that can tip the state of the system from the attractor of destructive conflict to the attractor of peace.
3. **Work the Edge of Chaos**: Mathematicians have identified moments in complex systems when they are particularly vulnerable to *phase transitions*; forks in the road that can lead to qualitatively different futures. It's useful to learn to read and recognize these edge-of-chaos states: Anomalies in systems—unusual incidents of tension, hardship, stress, aggression, violence, success, luck, or despair—may all be possible indicators of the presence of such states. These states usually represent *bifurcation points*: transition states between two possible states or attractors, so it is important to be familiar enough with the system to have a sense of what the latent and manifest alternatives might represent, and then work to tip the system in a more constructive direction. This can be done by searching for feasible, high-impact initiatives that might help trigger the transition from one state or attractor to another.
4. **Identify the Small Things that Change Everything**: In complex systems, many different factors interact to create attractors, but usually only a small subset of variables promotes really noteworthy changes in them. Even slight changes in these variables (called *control parameters*) can produce qualitative changes in the system's attractor landscape. These can take many forms, but the result is often dramatic changes in the conflict landscape. To translate this into a specific strategy for conflict resolution, it is necessary to identify the relevant

control parameters with the potential to change the attractor landscape constraining the behavior patterns. They are typically determined by the specific form and history of a conflict, and identifying them usually entails both an intimate knowledge of the situation and a bit of trial and error.

5. **Adapt to Change**: The Dynamical-systems approach to solving difficult problems is simply more open, flexible, and tolerant of ambiguity. It requires more reflection on *how* we are thinking and solving problems, as well as a keen recognition that data matters. It means recognizing that when feedback on the result of our actions comes in, that is the time to pay more attention, not less. And to make more decisions, not fewer. It involves starting wisely, making corrections in midcourse, and always learning from our mistakes. Because every situation is unique and circumstances are always changing, we must stay online in real time if we are going to continue to effectively navigate such problems. Especially when we think they are solved.

These practices are not tools or techniques for intervention; that would reflect linear cause-and-effect thinking. They are dialectical practices that incorporate many of the psychological principles of change, complexity, and coherence discussed in this text. Each is a way of thinking and acting that can enhance the capacity for addressing seemingly impossible conflicts and thereby increase the probabilities for constructive change. Each practice incorporates basic contradictions—*complexity/simplicity*, *creation/destruction*, and *change/stability*—that represent opposing human needs, tendencies, or processes. At their core, the practices involve an iterative process of managing such contradictions to change systems.

The approach enacted through these three practices aims to *transform the dynamics of the system maintaining the status quo of the conflict*. Specifically, its objectives are to regain a sense of *accuracy*, *agency* and *possibility* in the conflict; to achieve *sustainable solutions* by first opening up the closed system of the conflict to different information; and to then reconfigure the attractor landscape for the relationship.

From Theory to Practice

The dynamical approach to conflict has added value for understanding the nature of conflict and for providing broad guidelines concerning conflict management. In couching conflict in terms of basic processes that resonate with those observed and documented across otherwise quite distinct domains of human experience, our account meets the criteria of universality and parsimony that are basic benchmarks of science. Conflicts, however, are idiosyncratic in their manifestations, and this fact needs to be honored, not trivialized or ignored. To act on the basis of the basic principles and guidelines discussed in this chapter, strategies and tools need to be developed that can be implemented and assessed in "noisy" real-world contexts.

We have taken steps in this direction as well. This work is largely in its nascent stage, though, and thus likely to evolve in various (perhaps unpredictable) ways as we continue to translate the dynamical perspective into practice. But we feel sufficiently confident in what we have accomplished thus far to showcase two of our efforts. The first is a design for a series of meetings that apply the principles and practices of the dynamical systems approach to a multi-stakeholder conflict involving a violent and protracted labor-management-community dispute. The details of this design are provided in Appendix A. The second is a simulation program that can be employed in negotiations among stakeholders in a conflict. A full description of this program, as well as preliminary evidence regarding its effectiveness, is provided in Appendix B. Theses concrete translations of theory into practice focus attention on the potential of the dynamical approach for providing both synthetic understanding and a basis for concrete action—perhaps the most important of all benchmarks in science.

References

Axelrod, R., Riolo, R. L., & Cohen, M. D. (2002). Beyond geography: Cooperation with persistent links in the absence of clustered neighborhoods. *Personality and Social Psychology Review, 6*, 341–346.

Azar, E. E. (1990). *The management of protracted social conflict*. Aldershot: Dartmouth Publishing.

Bargh, J. A. (1996). Automaticity in social psychology. In E. T. Higgins & A. W. Kruglanski (Eds.), *Social psychology: Handbook of basic principles* (pp. 169–183). New York, NY: Guilford Press.

Barry, B., & Oliver, R. L. (1996). Affect in dyadic negotiation: A model and propositions. *Organizational Behavior and Human Decision Processes, 70*, 175–187.

Bennett, D. S. (1996). Security, bargaining, and the end of interstate rivalry. *International Studies Quarterly, 40*, 157–184.

Bercovitch, J. (2005). Mediation in the most resistant cases. In C. A. Crocker, F. O. Hampson, & P. Aall (Eds.), *Grasping the nettle: Analyzing cases of intractable conflict*. Washington, DC: United States Institute of Peace.

Blalock, H. M. (1989). *Power and conflict: Toward a general theory*. Newbury Park: Sage Publications.

Burton, J. (1987). *Resolving deep-rooted conflict: A handbook*. Lanham, MD: University Press of America.

Cacioppo, J. T., & Bernston, G. G. (1994). Relationship between attitudes and evaluative space: A critical review with emphasis on the separability of positive and negative substrates. *Psychological Bulletin, 15*, 401–423.

Coleman, P. T. (2003). Characteristics of protracted, intractable conflict: Towards the development of a meta-framework—I. *Peace and Conflict: Journal of Peace Psychology, 9*, 1–37.

Coleman, P. T. (2006). Conflict, complexity, and change: A meta-framework for addressing protracted, intractable conflicts—III. *Peace and Conflict: Journal of Peace Psychology, 12*, 325–348.

Coleman, P. T. (2011). *The five percent: Finding solutions to seemingly impossible conflicts*. New York, NY: PublicAffairs.

Coleman, P. T. (2012). The Essence of Peace? Toward a comprehensive and parsimonious model of sustainable peace. In P. T. Coleman & M. Deutsch (Eds.), *Psychological Contributions to Sustainable Peace*. New York: Springer.

Coleman, P. T., Bui-Wrzosinska, L., Vallacher, R. R., & Nowak, A. (2006). Protracted conflicts as dynamical systems. In A. K. Schneider & C. Honeyman (Eds.), *The negotiator's fieldbook: The desk reference for the experienced negotiator* (pp. 61–74). Chicago: American Bar Association Books.

Coleman, P. T., Fisher-Yoshida, B., Stover, M., Hacking, A., & Bartoli, A. (2008). Reconstructing ripeness II: Models and methods for fostering constructive stakeholder engagement across protracted divides. *Conflict Resolution Quarterly, 26*(1), 43–69.

Coleman, P. T., Goldman, J. S., & Kugler, K. (2009). Emotional intractability: Gender, anger, aggression and rumination in conflict. *International Journal of Conflict Management, 20*, 113–131.

Coleman, P. T., Vallacher, R. R., Bartoli, A., Nowak, A., & Bui-Wrzosinska, L. (2011). Navigating the landscape of conflict: Applications of dynamical systems theory to addressing protracted conflict. In D. Korppen, N. Ropers, & H. J. Giessmann (Eds.), *The non-linearity of peace processes: Theory and practice of systemic conflict transformation* (pp. 39–56). Opladen, Germany: Barbara Budrich Publishers.

Costa, P. T., & McCrae, R. R. (1995). Domains and facets: Hierarchical personality assessment using the revised NEO personality inventory. *Journal of Personality Assessment, 64*, 21–50.

Deihl, P., & Goertz, G. (2001). *War and Peace in International Rivalry: How do enduring rivalries between states affect international relations?* Ann Arbor: University of Michigan Press.

Klein, J., Goertz, G., & Diehl, P. F. (2006). The new rivalry data set: Procedures and patterns. *Journal of Peace Research, 43*, 331–348.

Deutsch, M., Coleman, P. T., & Marcus, E. C. (Eds.). (2006). *Handbook of conflict resolution: Theory and practice* (2nd ed.). San Francisco: Jossey-Bass.

Diehl, P. F., & Goertz, G. (2001). *War and peace in international rivalry*. Ann Arbor: University of Michigan Press.

Dijksterhuis, A., & Nordgren, L. F. (2006). A theory of unconscious thought. *Perspectives on Psychological Science, 1*, 95–109.

Dorner, D. (1996). *The logic of failure: Why things go wrong and what we can do to make them right*. New York, NY: Holt and Company.

Fisher, R. J., & Keashly, L. (1991). The potential complementarity of mediation and consultation within a contingency model of third party intervention. *Journal of Peace Research, 28*, 29–42.

Goertz, G., & Diehl, P. F. (1993). Enduring rivalries: Theoretical constructs and empirical patterns. *International Studies Quarterly, 37*, 147–171.

Gottman, J., Swanson, C., & Swanson, K. (2002). A general systems theory of marriage: Nonlinear difference equation modeling of marital interaction. *Personality and Social Psychology Review, 6*, 326–340.

Greenwald, A. G., & Banaji, M. R. (1995). Implicit social cognition: Attitudes, self-esteem, and stereotypes. *Psychological Review, 102*, 4–27.

Holland, J. H. (1995). *Emergence: From chaos to order*. Reading, MA: Addison-Wesley.

Kelso, J. A. S (1995). *Dynamic patterns: The self-organization of brain and behavior*. Cambridge, MA: MIT Press.

Kriesberg, L. (2005). Nature, dynamics, and phases of intractability. In C. A. Crocker, F. O. Hampson, & P. Aall (Eds.), *Grasping the nettle: Analyzing cases of intractable conflict*. Washington, DC: United States Institute of Peace.

Kuhn, T. S. (1970). *The structure of scientific revolutions* (2nd ed.). Chicago: University of Chicago Press.

Lewicki, R., Literer, A. J., Minton, J., & Saundars, D. M. (2004). *Negotiation* (3rd ed.). Boston: Irwin/McGraw-Hill.

Lewin, K. (1936). *Principles of topological psychology*. New York: McGraw-Hill.
Lewin, K. (1948). *Resolving social conflicts*. New York, NY: Harper & Brothers.
Marshall, M. G., & Gurr, T. R. (2005). *Peace and conflict*. College Park, MD: Center for International Development and Conflict Management.
McNulty, J. K., O'Mara, E. M., & Karney, B. R. (2008). Benevolent cognitions as a strategy of relationship maintenance: "Don't sweat the small stuff"...but it is not all small stuff. *Journal of Personality and Social Psychology, 94*, 631–646.
Murphy, S. T., & Zajonc, R. B. (1993). Affect, cognition, and awareness: Affective priming with optimal and suboptimal stimulus exposures. *Journal of Personality and Social Psychology, 64*, 723–739.
Nan, S. A., Mampilly, Z. C., & Bartoli, A. (Eds.). (2012). *Peacemaking: From practice to theory* (Vol. 2). Santa Barbara, CA: Praeger.
Nowak, A., Bui-Wrzosinska, L., Coleman, P. T., Vallacher, R. R., Bartkowski, W., & Jochemczyk, L. (2010). Seeking sustainable solutions: Using an attractor simulation platform for teaching multi-stakeholder negotiation in complex cases. *Negotiation Journal, 26*, 49–68.
Nowak, A., & Lewenstein, M. (1994). Dynamical systems: A tool for social psychology? In R. R. Vallacher & A. Nowak (Eds.), *Dynamical systems in social psychology* (pp. 17–53). San Diego: Academic.
Nowak, A., & Vallacher, R. R. (1998). *Dynamical social psychology*. New York, NY: Guilford Press.
Pearce, W. B., & Littlejohn, S. W. (1997). *Moral conflict: When social worlds collide*. Thousand Oaks, CA: Sage Publications.
Praszkier, R., Nowak, A., & Coleman, P. T. (2010). Social entrepreneurs and constructive change: The wisdom of circumventing conflict. *Peace and Conflict, 16*, 153–174.
Ruelle, D. (1989). *Elements of differentiable dynamics and bifurcation theory*. New York, NY: Academic.
Sandole, D. J. D. (1999). *Capturing the complexity of conflict: Dealing with violent ethnic conflicts in the post-cold War Era*. New York, NY: Pinter.
Schuster, H. G. (1984). *Deterministic chaos: An introduction*. Weinheim, Germany: Physik-Verlag.
Seligman, M. E. P., Steen, T. A., & Park, N. (2005). Positive psychology progress: Empirical validation of interventions. *American Psychologist, 60*, 410–421.
Staub, E. (2011). *Overcoming evil genocide, violent conflict, and terrorism*. New York: Oxford University Press.
Triandis, H. C. (1995). *Individualism and collectivism*. Boulder, CO: Westview Press.
Vallacher, R. R., Coleman, P. T., Nowak, A., & Bui-Wrzosinska, L. (2010). Dynamical foundations of intractable conflict: Introduction to the special issue. *Peace and Conflict: Journal of Peace Psychology, 16*, 113–125.
Vallacher, R. R., & Selz, K. (1991). Who's to blame? Action identification in allocating responsibility for alleged rape. *Social Cognition, 9*, 194–219.
Vallacher, R. R., & Wegner, D. M. (1987). What do people think they're doing? Action identification and human behavior. *Psychological Review, 94*, 3–15.
Vallacher, R. R., & Wegner, D. M. (2012). Action identification theory. In P. Van Lange, A. W. Kruglanski, & E. T. Higgins (Eds.), *Handbook of theories of social psychology* (pp. 327–348). Thousand Oaks, CA: Sage.
Von Bertalannfy, L. (1968). *General system theory*. New York, NY: Braziller.
Wegner, D. M. (1994). Ironic processes of mental control. *Psychological Review, 101*, 34–52.
Wegner, D. M., Vallacher, R. R., Macomber, G., Wood, R., & Arps, K. (1984). The emergence of action. *Journal of Personality and Social Psychology, 46*, 269–279.
Wilson, T. D. (2002). *Strangers to ourselves: Discovering the adaptive unconscious*. Cambridge, MA: Belknap Press of Harvard University Press.
Zartman, I. W., & Faure, G. O. (2005). *Escalation and negotiation in international conflicts* (1st ed.). New York, NY: Cambridge University Press.

Chapter 7
Sustainability: The Dynamics of Enduring Peace

In Chap. 1 we stated that the focus of this book is on conflict. However, all along we have had a hidden agenda (or latent attractor): as ultimately, we seek a better understanding and articulation of peace. Of course, peace is more than the absence of destructive conflict. Attempts to suppress conflict can, in fact, have the ironic effect of promoting catastrophic ruptures in interpersonal, intergroup, and international relations. Nor is peace a stable state. Rather, it evolves in the context of ongoing social relations. Sustainable peace must therefore be approached with an appreciation of dynamic processes, including the development and maintenance of attractors for social relations. Insight into the dynamics of sustainable peace can generate specific strategies for peacemaking and peacekeeping. Case studies of successful transformation from war to peace demonstrate the efficacy of these strategies.

In this chapter, we present a dynamical-systems approach to sustainable peace. We begin by defining peace in dynamical terms and outlining some of the many factors and conditions that can affect the probabilities of emerging patterns of stable peace versus destructive conflict. We then present a case study of the emergence of peace in Mozambique, which illustrates its non-linear dynamical properties and provides insights into some underlying mechanisms of peaceful attractors. We conclude with a sketch of our preliminary thinking on the development of a dynamical-minimal model of sustainable peace and a set of guidelines for its promotion.

Defining Peace

Specifying and defining peace has proven to be elusive; at one level, it is extremely complicated, but at another level, it is fundamentally simple. In a recent search of the Internet on articles published in English since 2000 with "peace" in their title, we identified over 40 terms distinguishing different types or aspects of peace (see Coleman, 2012). These include labels such as *agnostic peace*, *armed peace*,

R.R. Vallacher et al., *Attracted to Conflict: Dynamic Foundations of Destructive Social Relations*, Peace Psychology Book Series, DOI 10.1007/978-3-642-35280-5_7,

capitalist peace, *cold peace*, *democratic peace*, *hegemonic peace*, *feminist peace* and *fragile peace*. In fact, peace can differ in a wide variety of ways, including by level-of-analysis (from intrapersonal to interpersonal to international to global), direction (internally and externally peaceful), durability (from fragile to enduring), source (peace through coercion, democratic participation, economic incentive, etc.), quality (negative, positive and promotive peace), and scope (very local to global peace). This chapter is focused primarily on the dynamics of *sustainable peace*, also known as enduring, warm, lasting, durable, perpetual and dynamic peace.

But even the definitions of *sustainable peace* vary. For some, it simply means a stable, long-lived peace and is associated with the preservation of the status quo through adequate security and protection from outside influence. At the other extreme, sustainability is associated with adaptation and renewal—a creative-adaptive peace—which recognizes that all social systems are in flux and progress through multiple states or stages over time. This type of sustainable peace requires flexibility and responsiveness to important internal and external sources of change.

A basic challenge for most social systems is to perform both functions effectively: protect and preserve important aspects of the status quo while allowing for resilience and renewal when circumstances require (see Coleman, 2003; Siegel, 2010; Svyantek & Brown, 2000). This tension captures our view of sustainable peace as a constructive manner of relating to one's self, others and the environment that is both stable and dynamic, resulting in a process that provides a secure sense of integrity and stability, and an ability to adapt to change when necessary or useful. From this perspective, sustainable peace is a process rather than state. Sustainability relates to the properties of this process.

Although important differences may exist between different types and levels of sustainable peace (such as dyadic, intergroup, and international forms), they all share some basic qualities reflecting a relative absence of destructive tension, conflict, and violence, and a relative presence of constructive conflict, harmony and well-being. Accordingly, we define sustainable peace as existing in *a state where the probability of using destructive means or violence to solve problems is so low that it does not enter into any party's strategy, while the probability of using cooperation and dialogue to promote justice and well-being is so high that it governs social organization and life* (Coleman, 2012).

In dynamical terms, sustainable peace characterizes a system in which social processes evolve into, and stabilize in the regions of a state-space corresponding to constructive interactions, and far from regions associated with destructive interactions such as violence and injustice. The probability of peace can then be defined as the likelihood that a system's processes are contained in this region of the state-space. The probability of destructive conflict is therefore the likelihood and frequency of system's processes visiting the region of the state-space corresponding to destructive interactions and violence.

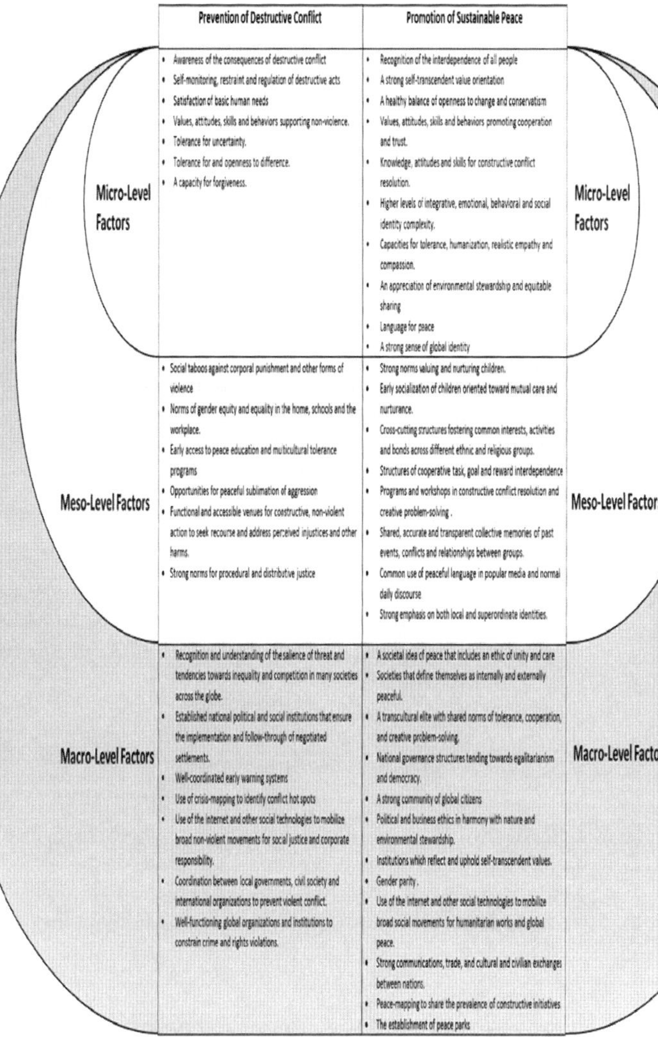

Fig. 7.1 A nested model of components of sustainable peace (Coleman, 2012)

Addressing the Probabilities of Peace

In the book, *Psychological Components of Sustainable Peace* (Coleman & Deutsch, 2012), a group of distinguished scholar-practitioners of peace and conflict provide a review of empirical research on a wide variety of factors associated with peaceful individuals, groups, and societies. In its conclusion, the authors present a nested-model of sustainable peace (Coleman, 2012; see Fig. 7.1), which organizes many of these factors by level (micro, meso, and macro) and by orientation (prevention of destructive conflict or promotion of sustainable peace). Given our dynamical view of sustainable peace, the many factors, conditions, and processes associated with it can be understood in the context of their relative effects on (1) decreasing probabilities for moving and maintaining a system's processes in the direction of a region of injustice, destructive conflict and violence, and (2) increasing probabilities for moving and maintaining a system's processes in the direction of a region corresponding to fairness, cooperation and lasting peace.

The factors associated with preventing destructive conflict can operate in three distinct ways. Some prevent a buildup of social tensions that can result in destructive means of recourse. Others affect the capacity of social systems to self-regulate—that is, to detect and modify destructive tendencies and violence when they do arise. The third group of factors can facilitate the escape of social systems from vicious cycles of destructive escalation and reprisal. The three groups of factors operate at the micro, meso, and the macro levels of social reality, but may take a different form at each level.

Other factors have been described as promoting sustainable peace. Although these factors may also take different forms at different levels of social reality, they typically share a common denominator. They all address different aspects of the capacity of individuals, groups, and societies to understand themselves as part of a larger system, and to work for the benefit of the entire system rather than only for the particular subcomponent to which they belong and which they see as in competition and conflict with the rest of the system.

Here a few of the factors are summarized in brief.

Micro-level Factors for Prevention of Destructive Conflict

Destructive conflict is most likely to be mitigated when individuals in a group or society share the following components:

- Satisfaction of basic human needs. Including physiological needs, safety, and dignity. Frustrated basic human needs increase the tendency for aggressiveness and violence. If basic needs are satisfied, there is a lower propensity for destructive interactions in social systems, so social processes naturally evolve in a more cooperative way.
- Awareness of the causes, consequences, and escalatory tendencies of destructive conflict and violence. Sufficient levels of consciousness of the deleterious effects of destructive conflict on relations and resources typically act as an inhibiting force, thus decreasing probabilities of individuals instigating or resorting to such means.
- High levels of tolerance for, and openness to difference. A press for coherence is one of the central tendencies of psychological and social systems. The presence of differences among elements of a system evokes mechanisms that function to reduce the differences and restore coherence. A proportion of these mechanisms

rely on coercion, aggression, and violence against the alien entities. The higher the level of tolerance for difference, the lower the likelihood that differences will evoke aggressive responses.

- Moderate levels of tolerance for uncertainty. Uncertainty stimulates fear and a tendency to seek safe, stable, and familiar states and solutions. Uncertainty thus stimulates a need for control and for the suppression of members of outgroups, who can increase ambiguity and unpredictability in social systems.
- Moderately high levels of self-monitoring, restraint, and regulation of internal impulses for destructive or violent acts. Research has identified the critical link between self-monitoring and self-regulation and tendencies to constrain impulses to react destructively to situations before consideration of the consequences (Mischel, DeSmit, & Kross, 2006). Such internal forms of inhibiting feedback reduce the likelihood of escalatory spirals being triggered and increasing hostilities.
- Values, attitudes, and skills supporting non-violence. Philosophies, strategies and skills for acting strategically through non-violent methods provide a means of recourse for grievances that allow individuals and groups to address their concerns actively and effectively, thus diminishing the need or tendency to respond to wrongs with aggression or violence (Gurr, 2000).
- A capacity for forgiveness. When a transgression has occurred and the perpetrator has acknowledged the act and expressed regret, a capacity to forgive and let go of the resentment and negativity associated with the act is critical to escaping reinforcing feedback loops of retribution and escalation, and thus, decreases the odds of bolstering more destructive regions of the state space.

Micro-level Factors for Promotion of Sustainable Peace

Individuals can also increase the probabilities for sustainable peace when they adopt and internalize these factors:

- Recognition of the interdependence of all people, similar and different, local and global. Decades of empirical research has demonstrated the fundamental importance of attitudes and perceptions of positive interdependence between people (We sink or swim together) on constructive conflict and group dynamics at the interpersonal, intergroup, and international levels (Deutsch, 1973, 2006).
- A strong self-transcendent value orientation committed to the welfare of others and society. Schwartz (1992) proposes an integrated system of ten universal values of social systems organized by two bi-polar dimensions: Self-transcendence to self-enhancement and openness to change to conservatism. Individual's with more self-transcendent values tend to act more cooperatively and peacefully, while those with more self-enhancement values tend to be more inclined to be authoritarian, and to bully and evidence violent behavior (see Fry & Miklikowska, 2012).

- An optimal balance of openness to change *and* conservatism, responsive to changing times and circumstances. Openness to change and conservatism have likewise been pitted against one another in the literature (and political life!). Conservative values such as conformity, tradition, and security are often associated with less tolerant and more rigid attitudes and behaviors, while more open values are not. However, we suggest that an optimal balance of both orientations is more likely to move social systems into regions of a state-space that are both stable and adaptive, thereby fostering a more sustainable form of peace.
- Skills and behaviors promoting cooperation and trust. Johnson and Johnson (1989, 2005) have written extensively on the importance of developing, in addition to cooperative values and attitudes, the appropriate skills for enacting cooperation in social relations. These include active listening skills, group process facilitation, respectful discourse, and flexibility in thought and action.
- Knowledge, attitudes, and skills for constructive conflict resolution. In addition to skills conducive to cooperation, skills for managing conflicts constructively when they do arise are also critical. These include being able to distinguish positions from needs, reframe conflicts from competitive struggles to mutual problems, creatively fashion integrative or complementary solutions, and establish fair criteria for agreements (Deutsch, 2006; Fisher, Ury, & Patton 1991).
- Higher levels of cognitive, emotional, behavioral, and social identity complexity. Consistently, research on different forms of individual-level complexity (cognitive, emotional, behavioral and social identity; (see Koo & Kim, 1999; Koo, Han, & Kim, 2002; Liht, Suedfeld, & Krawczyk, 2005; Roccas & Brewer, 2002; Winter, 2007) has found positive associations between high-complexity and more positive social experiences and relations. This has been true particularly in situations of conflict, where the pull toward over-simplification can become great (see Kugler, Coleman, & Fuchs, 2011, Working paper; Seudfeld, Tetlock, & Winter, 2001).
- Capacities for tolerance and realistic empathy (understanding how a situation looks to someone else). It is unrealistic that people will manage to have cooperative, harmonious relationships with everyone they encounter. Nevertheless, the capacity to tolerate and take the point-of-view of others, even those with whom we have stark value differences or a contentious history, goes a long way in increasing the probabilities of peaceful relations.
- Language for peace: a large lexicon for all aspects of cooperative and peaceful relations and sufficient use of such terms to foster automaticity. Gomes (2012) has elaborated on the critical role language plays in helping construct a vision and a social-linguistic environment for constructive conflict and peaceful relations.
- A strong sense of global identity with a concrete understanding of the steps that need to be taken locally to act as a global citizen. Research has long established the value of joint-superordinate goals for ameliorating intergroup tensions. Similarly, shared, meaningful identities between members of different groups

and nations set the stage for mutual problem-solving and increased compassion (see Deutsch, 2012; Lindner, 2012).

- An appreciation of environmental stewardship and equitable sharing of the Earth's resources among its members and with all human beings. Related to a global sense of identity is the nature of each individual's relationship to the environment and the planet. A heightened awareness of the interdependence among people and the environment fosters a more transcendent orientation and makes salient the need for multi-stakeholder collaboration (Deutsch, 2012).

Meso-level Factors for Prevention of Destructive Conflict

Some of the many meso-level factors relevant to conflict prevention and mitigation identified by scholars include:

- Social taboos against corporal punishment and other forms of violence in the home, schools, workplace, and public spaces. Research on peaceful societies across cultures has shown that communities that are more peaceful internally among their own members and externally with members of different communities tend to evidence strong social norms and taboos which prohibit the use of violence in many forms (Fry, 2006). In dynamical terms, these act as repellers for destructive interactions.
- Norms of gender equity and equality in the home, schools, and the workplace. Research indicates that groups and societies that maintain equal and equitable relations between men and women also tend to evidence less aggression (Disney & Gbowee, 2012; Druckman & Alban, 2012).
- Early access to peace education and multicultural tolerance programs in preschool, elementary, and middle school. When schools and communities provide early exposure to tolerant attitudes and effective conflict management skills, the effects trickle up, eventually impacting emergent social norms and more peaceful climates (Johnson & Johnson, 2005; Sandy & Boardman, 2000).
- Opportunities for peaceful sublimation of aggression through competitive or extreme sports, occupations, creative arts, etc. These opportunities, particularly for urban youth and others who might find it more difficult to gain access to expressive outlets, can provide a cathartic and much needed venue for ventilation of pent up frustration and aggression (Gratch, 2012).
- Functional and accessible venues for constructive, non-violent action to seek recourse and address perceived injustices and other harms. A key community structure that has been found to help address experiences of frustration from deprivation and reduce organized political violence represent fair and effectively functioning methods for achieving recourse for grievances suffered (Gurr, 2000).
- Strong norms for procedural and distributive justice in schools, workplaces, marketplaces, and elsewhere in the community. Similarly, when policies and

practices for fair intercourse are widespread in communities, they can have a mutually reinforcing effect, bolstering a strong climate for fairness and justice (Deutsch, 2006).

Meso-level Factors for Promotion of Sustainable Peace

Peace scholars have also identified the following promotive mid-level components as critical:

- Strong norms valuing and nurturing children. Anthropological research has found a significant positive relationship between warm and caring norms that value and nurture children and environments replete with more constructive and respectful adult interactions (Fry, 2006).
- Early socialization of children oriented toward mutual care and nurturance. In addition to teaching skills to manage conflict, schools that model and support cooperation and teamwork among students help to shape the skills and attitudes conducive to more harmonious adult relations (Johnson & Johnson, 2005).
- Cross-cutting structures fostering common interests, activities, and bonds across different ethnic and religious groups. One of the most important findings from ethnographic research on violent versus peaceful communities is the value of cross-cutting structures (multi-ethnic sports teams, labor unions, political parties, etc.) for connecting members of different ethnic groups, building relationships and mitigating escalation of conflict when it occurs (Levine & Cambell, 1972; Varshnay, 2002).
- Structures of cooperative task, goal, and reward interdependence in schools, workplaces, and politics. The attitudes and skills for cooperative interdependence described at the micro level are typically induced, developed, and maintained by task, goal, and reward structures that incentivize working together (Johnson, 2012).
- Programs and workshops in constructive conflict resolution and creative problem-solving for children, adults, parents, and leaders of schools, businesses, and politics. The positive side of constructive management of conflict is that it can lead to creative solutions, innovation, and breakthroughs to mutual problems, thus increasing positivity in the system and reinforcing the value of constructive conflict management (Johnson, 2012).
- Shared, accurate, and transparent collective memories of past events, conflicts and relationships between groups. Protracted conflicts are often maintained in part by contrasting memories of past events, heroes, and villains associated with the disputes that are held by the opposing parties. However, when more balanced and accurate accounts of such histories are shared, the odds of a community healing and moving on after conflict increase considerably (Nets, 2012).

Macro-level Factors for Prevention of Destructive Conflict

The following macro-level factors have been found to be important to the prevention and mitigation of destructive conflict.

- Establishing national political and social institutions that ensure the implementation and follow-through of negotiated settlements. Peace treaties between previously warring parties are essentially worthless and short-lived unless there are stable institutions responsible for their effective implementation (Diehl & Druckman, 2011). These are critical structures for preventing relapse into violence.
- Well-coordinated early warning systems operating through local governments and NGOs networked locally, regionally, and globally for efficient communication. The presence of accurate and effective early-warning systems for identifying and communicating trends and increases in hostilities between ethnic groups can provide much-needed inhibiting feedback containing destructiveness if officials' responses to such information are swift and appropriate.
- Use of crisis-mapping: an open-source platform for collecting and plotting local cell-phone accounts of the commission of violent atrocities to inform the international community of emerging crises in a timely manner. During the political violence in Kenya in 2008, individuals outside the country organized local people to use their cell phones to photograph, document, and report incidents of violence, thus providing the international community with information that could be used to confront officials who denied such incidents. Again, such capacities decrease the likelihood that officials will employ more flagrant forms of coercion or attempt to deny it when they do.
- Use of the internet and other social technologies to mobilize broad non-violent movements for social justice and corporate responsibility. This method has been used with increasing frequency by young people and other activists organizing politically in the Middle East and demanding social justice during the "Occupy" movement. Such tools help to keep in check flagrant abuses of power by those in authority through means of civil disobedience.
- Coordination between local governments, civil society, and international organizations to prevent violent conflict. Unfortunately, this is a conflict-mitigating factor rarely realized today. However, such cross-sector coordination holds great promise in the not-so-distant future for helping to prevent conflict escalation and to address it more effectively when it occurs.
- Well-functioning global organizations and institutions such as the United Nations, the International Criminal Courts, Interpol, and the Universal Declaration of Human Rights. Despite their flaws, international institutions like these signal the potential consequences to war criminals and others committing atrocities in the name of war and, under certain circumstances, can inhibit such policies and acts.

Macro-level Factors for Promotion of Sustainable Peace

Finally, these factors have been found to be highly influential in promoting and maintaining a state of sustainable peace at a macro-level:

- A societal idea of peace. Groups and nations are much more likely to evolve in peaceful directions if they have a clearly specified sense of what this entails. Such visions must include an ethic of interethnic unity, care and nurturance of others, that is as strong as (or stronger than) the view of peace as something that needs be secured and defended. The *Universal Declaration of Human Rights* is an excellent starting point.
- Societies that define themselves as internally and externally peaceful. Research has found that when societies define themselves as peaceful, they are much more likely to behave and organize themselves in a consistent manner (Fry, 2006). Today, Iceland, Denmark, Canada and Norway provide good examples.
- A transcultural elite with shared norms of tolerance, cooperation, and creative problem-solving, who model for all the efficacy and value of constructive, non-violent action. The rise of such an elite class of individuals (particularly leaders of business, government, celebrities and professional athletes), signals to the broader population the value and importance of behaving in a compassionate and self-transcendent manner (Deutsch, 2012).
- Gender parity with a proportional number of women in the highest positions of leadership in business, politics, and the military. Women continue to be widely under-represented in many positions of leadership, particularly those involving peace processes.
- National governance structures tending towards egalitarianism and democracy. Research has also shown that democracies rarely go to war with other democracies.
- Political and business ethics that are in harmony with nature and environmental stewardship.
- Institutions which reflect and uphold self-transcendent values. As Fry and Miklikowska (2012) have written, when institutions model, incentivize, and support such goals and values, the likelihood of them being adopted and internalized increases exponentially.
- Use of the Internet and other social technologies to mobilize broad social movements for humanitarian works and global peace. As discussed above, social technologies are today used to mitigate and prevent violence. However, such tools also offer great promise for coordination and support for peace building and other prosocial initiatives as well.
- Strong initiatives for communications, trade, and cultural and civilian exchanges between nations. Such structures simply establish cross-cutting structures and positive interdependence between nations.

Together, these factors can operate in concert to constitute a *system of sustainable peace*—in other words, societal attractors for peaceful relations—

distinguishing such societies from those locked in *systems of destructive conflict*. As other dynamical research has shown (see Gottman, Murray, Swanson, Tyson, & Swanson, 2002; Kugler et al., 2011, Losada, 1999), when the ratio of positive dynamics to negative dynamics in social systems is high (somewhere between 3:1 and 5:1), the odds of healthier, thriving relations and societies tends to increase markedly. When the positivity/negativity ratio is closer to 1:1 or lower, the relatively stronger effects of negativity will drive the system into either a sustained state of hostilities or lead to its disintegration.

As is evident from the rather lengthy list of factors outlined above, sustainable peace, at this level, is complex and multiply determined. The good news is this provides us with many venues through which to increase positivity and contain negativity. However, this degree of complexity also leaves us with a rather fuzzy, imprecise understanding of the pathways to peace. Below, we discuss our attempts to drill down more deeply to identify a smaller set of more basic parameters that might account for sustainably peaceful relations. However, first we discuss when families, communities, and societies do fall into deep and wide attractors for destructive conflict. Here, it becomes critical to understand how change unfolds in these settings, and how peace may return and stabilize. Below, we offer a summary of a case study of the emergence of peace from a protracted conflict in Mozambique (see Bartoli, Bui-Wrzosinska, & Nowak, 2010; Coleman, Vallacher, Nowak, Bui-Wrzosinska, & Bartoli, 2009).

The Dynamics of Peace in Mozambique

Mozambique has been at peace, independent and united since 1992. While independence from Portugal was achieved much earlier (June 25, 1975), the country was not able to enjoy the newly gained independence due to an internal war that challenged its peace and unity. Mozambicans experienced a protracted, intractable conflict due to complex dynamics that can be better analyzed now, at a distance. United Nations officials have recognized the case as a success and its analysis is offering fresh insights into our collective understanding of peace and conflict.

After more than 400 years of colonial ruling by the Portuguese, Mozambique acquired independence suddenly when a military coup in Lisbon ended military control of the colonies. Years of independence struggle by the Frente de Liberacao de Mozambique (FRELIMO) were crowned by success and a new government was formed. How did it happen that the Mozambicans, after such a success, soon entered into a protracted internal war? We believe that a process of unresponsiveness of the government to social feedback led to an internal clash, profoundly dividing the whole nation.

Ideologically, FRELIMO was Marxist-Leninist and interpreted the initial joyous, festive reactions of the population to the newly gained independence as support of their socialist policies (abolition of private property; control of movement; massive indoctrination). When a growing part of the population (including former local leaders) tried to express their dissent they were labeled as counterrevolutionaries and isolated. This negative response fueled an even more extreme response among the dissenters until violence was chosen as the 'only' remaining option. White-supremacist Rhodesia took advantage of these internal dynamics to fuel the divisions further by supporting and arming a nationalist fraction, the Resistencia National de Mozambique (RENAMO). This support was another reason for the socialist government of Mozambique renewing efforts to win the conflict militarily. The hardening of positions between the two parties (FRELIMO and RENAMO) continued in a spiral that soon became self-reinforcing and capable of co-opting all resources and channels of influence (including the regional, geopolitical ones) available to the parties into a destructive, confrontational mode.

During the period from 1977 to 1992, the parties to the conflict created an enmity system characterized by exclusionary dynamics that were intentionally executed to diminish the enemy and 'win the war'. From the recruitment strategies of RENAMO (including the kidnapping and exploitation of child-soldiers) to their effort to intentionally target the existing infrastructures, it soon became clear that the main objective of the war was to eradicate the adversary and introduce chaos that would make governance simply impossible.

Yet, from July 8, 1990 until October 4, 1992, the Mozambicans were able to launch and execute a successful peace process, which continued years later with the intervention of the United Nations and subsequently confirmed by a series of democratic elections. How could the Mozambicans move out of this system of enmity and how is a dynamical system approach helpful in understanding the transition?

Mozambicans went through a long process of conflict escalation, maintenance, and transformation. They created an enmity system into which all parties were trapped, and soon became the victims of their own conflict strategies. The perspective of dynamical systems helps us to see the dynamic forces at play that were maintaining and perpetuating a state of intractable conflict: how one actor countered the move of the other, how the conflict dynamics at place exhibited generative qualities, and how entrapment was operative through identified recurrent feedback loops in the conflict system.

It can be argued that peace—as a respectful, mutually recognizing and empowering relational system—was trying to emerge long before the actual peace process. However, the relational system between the two main contenders was so strong and rigid that neither party could provide space for the other to fully or properly express their views. The very presence of the other party was perceived as a threat and the possibility of an agreement with the enemy was viewed as defeat. Parties to the conflict were trapped into a strategy of mutual destruction that made it not only impossible for the enemy to function as a governing force, but also for anyone else to truly and legitimately serve the population politically. For many years, the parties were confined to a situation were they refused to believe that it was possible to create an inclusive Mozambique. Yet, they were unable to achieve the idealised, exclusive version of it through a definitive, partisan victory. They were deadlocked.

The first glimpses of peace arose in people that were close enough to the centers of power to recognize that the current strategies were not working, but, at the same time, were free enough from the constraints of the enmity system. These catalysts, actors that made 'the new next move' possible, played a particularly important role in the successful transition of Mozambique from war to peace. The enmity system, restricted the relational space, made the exploration of constructive alternatives impossible. However, a handful of individuals who were dissatisfied with the current oppositional dynamics, such as Jamie Pedro Goncalves, Catholic Archbishop of Beira, made the emergence of new options in such a frozen system possible. He is the one who allowed the inclusion of Alfonso Dhlakama, the leader of RENAMO—a key figure in the conflict landscape who was utterly damned by the government and the international community—into what later became a political debate.

The first meeting of Alfonso Dhlakama and the Archibishop was organized secretly and was marked by enormous risks: RENAMO was feared for their cruelty and violence, and meetings could only take place in the hidden headquarters of RENAMO. Yet, this first face-to-face encounter in the bush proved to be groundbreaking, and allowed for an exploration of options that included a negotiated settlement. An important factor in the success of this first encounter was serendipity: Goncalves and Dhlakama happened to share the same, rare, native language, which enabled them to communicate freely and properly. This experience of direct, un-constrained communication, enriched and deepened by the experience of their common roots, was the basis for the subsequent years of political work that transformed RENAMO into a recognized political party.

As a catalyst, Goncalves initiated the process of conflict transformation as he was able to internally reconcile contradictory characteristics that he shared with both sides in the conflict. Whereas most bishops in Mozambique, at that time, were White, he was one of the first native, black bishops, nominated by the Catholic Church. Despite being a Catholic (which the FRELIMO government associated with the colonialist Portuguese), he had established close ties with the Communist FRELIMO party, and he had been engaged in initiatives aimed at easing tensions between religious communities and the government. Goncalves had a rich education, both locally and abroad, so he had been exposed to various cultural and social environments, but, at the same time, he was able to maintain a strong sense of his local roots. He was fluent in various ethnic languages and was able to nurture a vast social network within the community of Beira, the second largest city in Mozambique.

Goncalves' actions as a mediator were endorsed by key figures since he was highly valued for his abilities to listen to contradictory positions without taking sides and to creatively approach existing conflicts. This positioned him at the center of a constant flow of information within and between various stakeholders, which allowed him to consider options that could not have been envisioned by others. He had the capacity to deeply resonate with the cultural and political categories relevant to each side of the conflict, by appreciating FRELIMO's values and objectives, as well as RENAMO's central themes that would have been otherwise obscured.

But foremost to the peace process were his close contacts with the Community of Sant'Egidio, a Catholic NGO that later proved central in the mediation process between FRELIMO and RENAMO that eventually brought peace to the country. His particular position allowed for a realistic, informed and strategic use of his social network as well as local and international resources during the peace process. Goncalves played a central role in the local society: many people shared their experience of the conflict with him, which made him a focal point—or in network terms, a *connector* (Barabasi, 2001). This, in turn, incrementally developed his desire for a solution: the more profound his understanding of people's different perspectives on the conflict, the more he felt committed to act proactively to find peace. Nevertheless, his approach was inclusive and empowering from the beginning: instead of becoming an advocate for an *a priori* solution, he started exploring the possibilities for peace emerging from the parties themselves. He welcomed new ideas and transmitted them through often non-existent communication channels, facilitating and establishing contact between the parties.

From a DST perspective, the impact of a catalyst like Goncalves on highly dense and extremely polarized conflict systems can be explained with reference to two basic mechanisms. First, a catalyst introduces discontinuity into tightly coupled, highly coherent, but polarized communication networks. Second, a catalyst weakens polarizing tensions by constructing intermediary structures within himself or herself first, where two exclusive states are reconciled internally. Because of these mechanisms, an integrative perspective on the issues inherent in the conflict can transcend the divisions imposed externally, by the conflict. The first mechanism disrupts the enmity system, which consists of a polarized social network with strong, reinforcing links and loops among nodes (members) within each side of the conflict, and strong, inhibiting links and loops between nodes (members) of the opposing parties to a conflict. This structure is very stable, since both types of links exert pressure on each element of the network to maintain their respective enmity states (opinions, positions on conflict issues, motives), which are similar within each group, and, at the same time, opposed to the other group—thus, any change of the state of one element will be immediately counteracted by these joint forces. Since a catalyst is connected by positive, reinforcing links to both sides of the conflict, he/she introduces a discontinuity into the system by disrupting the strong internal coherence and unsettling the balance of forces preserving the system's *status quo*.

The integrative mechanism introduced by the catalyst leads to depolarization through the formation of an internal, mental representation in which exclusive elements of the conflict are integrated into an intermediate structure somewhere between the seemingly incompatible mindsets of the parties. This is not an obvious or easy process, as it goes against natural, social psychological processes. In fact, research shows that attitudes on important issues tend to be either highly positive or highly negative (Latane' & Nowak, 1994). Intermediary states, although logically possible, are psychologically unlikely. Indeed, when judgements are important, ambivalence is manifested as oscillation between two opposing judgments (Nowak & Vallacher, 1998; Vallacher & Nowak, 1994a, 1994b). Nonetheless, integration of ambivalent or exclusive states may be possible through the establishment of coherent emotional-cognitive structures. Individuals who do come up with an integrative solution are usually full of contradictions because the conflict dividing the society has often been internalized (with respect to their emotions, identity, cognitive structures, or worldviews). In this context, endogenous actors in a conflict may be far more effective than mediators from the outside, since they have the capacity to actively participate emotionally and cognitively in such a process, and only from such a position can they integrate a maximum of relevant elements polarizing the conflict (Bartoli, 2005).

This may contribute to the understanding of why outside interventions often fail in their attempts to resolve intractable conflicts (Kriesberg, 2005). Initially, only few individuals can hold the ambiguity of seemingly irreconcilable conditions in their mind. However, this view can spread, initiating a gradual process of dissolution of the enmity system. The integrated perspective, as well as the unique position of the catalyst in the social networks of the conflict, allow for the spread of novel solutions and ideas. We assume that peace emerges when what is conceptualized in the mind of catalysts manifests itself socially through the relationships with others in the network.

In summary, the dynamics of the movement from a system of war toward a system of peace in Mozambique can be characterized as a movement along a primary dimension: From closed, ultra-coherent, maladaptive dynamics toward more open, complex, adaptive dynamics. The conflict system, governed by the logic of war, was unresponsive to local needs and demands. In a sense, many regional and global issues (Apartheid, The Cold War) were initially "coupled" with the conflict dynamics within the country. The peace process changed the nature of the country's dynamics: Mozambique was "decoupled" from maladaptive global and regional processes, and started to respond to local challenges, demands, and needs of the population and the elite. This is visible when we look closer at the nature of the solutions and strategies applied to specific problems of the post-war reality. This is strongly related to the process of "regaining freedom of movement" of individuals from the constraints of the enmity system.

We believe that the description of the peace process in Mozambique in dynamical terms provides a framework within which subtle, micro, but qualitatively important events can be directly connected to the macro-level peace processes they triggered. This contributes to a better understanding of the process of creation and later transformation of conflict intractability, linking the properties of intractable conflicts to corresponding transformation strategies and to the emergence of peace.

A Dynamical-Minimal Model of Sustainable Peace

As the dynamical-systems approach suggests, all communities and societies affected by destructive conflict, such as war in Mozambique, have a latent potential for peace, and societies at peace often harbor a latent potential for hostilities. For instance, in *Islands of Agreement: Managing Enduring Armed Rivalries* (2007), Gabriella Blum describes the many instances of cooperation operating in the context of long-enduring armed rivalries such as between India and Pakistan, Greece and Turkey, and Israel and Lebanon. These pockets of cooperation reduce suffering and loss and allow mutually beneficial exchanges to take place, and are evidence of the latent potential for peace inherent in all societies, even those currently trapped by war. On the other hand, if we examine the state of the peace process in Northern Ireland today, we see a somewhat fragile state of peace and often hear the rumblings of what could once again become a manifest dynamic of violence. Both potentials coexist. The objective in peace processes is not only to move from one state (war) to the other (peace) constructively. The priority today should be on sustainability—how to increase the probabilities that once societies transition to peace, they will be able to remain there and navigate the many challenges to peace that can accompany its implementation and maintenance.

A theoretical approach to understanding and promoting sustainable peace is currently being developed, informed by the efforts of our multidisciplinary research team working to apply insights and methods from complexity science (see Coleman, Bui-Wrzosinska, Vallacher, & Nowak, 2006; Coleman, Hacking, Stover, Fisher-Yoshida, & Nowak, 2008; Coleman, Vallacher, Nowak, & Bui-Wrzosinska, 2007; Coleman, Vallacher, Nowak, Bui-Wrzosinska, & Bartoli, 2011; Nowak, Vallacher, Bui-Wrzosinska, & Coleman, 2006; Vallacher, Coleman, Nowak, & Bui-Wrzosinska, 2010). We suggest that qualitative differences in the dominant patterns of social behavior (such as those found in peaceful societies versus hostile or warring societies) can be accounted for at a very basic level by relatively few factors. Accordingly, our research has been attempting to identify, from scholarship and practice, the fundamental factors that promote sustained peaceful dynamics in communities or, put another way, that make societies immune to prolonged destructive or violent conflict.

Our ultimate goal is to identify the most basic parameters that constitute the dynamics of sustainable systems of social peace. To date, our empirical research and those of others modeling social dynamics (Gottman, 1993; Gottman, Murray, et al., 2002; Gottman, Swanson, & Murray, 1999; Gottman, Swanson, & Swanson, 2002; Losada, 1999; Losada & Heaphy, 2004) have identified a small set of parameters which are known to affect qualitative differences in social systems. These include:

- **Flexible structures**: These enable a system to operate in a manner that provides a secure sense of integrity and stability as well as an ability to adapt to change (Siegel, 2010).

- **Optimal adaptation-integration ratio**: Related to the above, this describes the ratio of time a social system spends in an integration mode (providing a sense of stability and clarity of purpose) and an adaptive mode (providing sufficient openness and receptiveness to critical changes in the environment (see Svyantek & Brown, 2000).
- **High complexity dynamics** (cognitive, emotional, behavioral): Such dynamics are common for individuals, dyads, and groups who are able to navigate conflicts constructively and solve problems creatively (Gottman, Murray, et al., 2002; Kugler et al., 2011; Losada, 1999; Suedfeld et al., 1992; Tetlock, 1985).
- **High positivity-negativity ratio** (emotional pooling): This is central to positive lasting social relations (Gottman, Murray, et al., 2002; Kugler et al., 2011; Losada, 1999).

From the dynamical point of view, a society characterized by sustainable peace is one in which the attractor for positive interactions is strong—that is, with a deep and wide basin of attraction—and the attractor for violent interactions is practically non-existent. Because single, even grave, acts of violence can happen in practically any society, the absence of an attractor for violence in practical terms means that the basin of attraction of the positive attractor extends even to very violent acts. In other words, even very violent incidents will eventually end up resulting in a peaceful process of repair and reconciliation.

The strength of a positive attractor is directly related to the richness of processes of positive interactions in the society. High social and cultural capital—that is, rich networks of social relations characterized by trust, the existence of many cross-ethnic groups and organizations to which individuals belong, strong culture, rich arts—all contribute to the strength of the positive attractor. The existence of organizations and institutions that can mediate individual expressions of needs and influence the system also extends the basin of attraction of the positive attractor. If there exists an institution that can help individuals to attain their goals, express their needs and grievances, individuals will likely use this institution if they feel dissatisfied with the current state of affairs. If they do not perceive this as an option, they are likely to take things in their hands and resort to direct, even violent action (Gurr, 2000).

The capacity of social systems to prevent the buildup of tensions is related to their capacity for movement. Moreover, because any system exists in the context of a constantly changing external environment, and because there is a constant buildup of tensions due to processes occurring at the level of elements (i.e., changes in situations of individuals, emergence of new needs, changes of goals and values, etc.), constant movement and change is required for adaptation. If such movement is constrained, tensions build up and can promote the development and strengthening of an attractor for violent solutions. In other words, a social system needs to change and evolve to remain functional. It needs constant adjustments to dissipate tensions. So, if all the elements (e.g., social groups) can have some influence on decisions, or at least opportunities of expressions of grievances, the system evolves in small steps releasing the tension. If, however, the capacity for movement (constant adjustments) is blocked and the repressed tension creates a strong negative attractor, at some point, there will be sudden change as the system flips into the attractor for violence. All the factors that allow for nonviolent expression and participation contribute to sustainable peace. From this perspective, constructive conflict, which can address tensions without resorting to violence, contributes to sustainable peace.

An important factor in avoiding the buildup of tension is the perspective adopted by individuals, social groups, and organizations. A more self-centered perspective, oriented toward the benefit of the element of the social system alone, is likely to create tensions between elements, since individual elements are likely to have divergent interests. A more global perspective, taking into account the benefits of the whole system, is likely to result in a much more coherent direction of thought and action among elements, therefore creating less tension. Because elements of the system on their own, acting in a bottom up manner, are unlikely to resolve all the problems and avoid all the tensions, it is important that there are groups and institutions that can identify the rising potential for violence and modify the societal dynamics in such a way that the social tension is reduced. Deconstructing the negative attractor represents an important dimension of this process.

As discussed in the section on the Crude Law of social relations, competitive interactions, especially when they are protracted, create competitive social orientations and a proclivity for further violence. In terms of dynamical systems' thinking, competitive interactions create and deepen attractors for violent interactions. Societies characterized by sustainable peace have the ability to avoid acts of violence and to escape from cycles of violence if violence occurs. Thus, social norms and values against violence, as well as the presence of institutions and organizations, both formal and informal, that can prevent violence, and break the cycle of violence, represent essential components of sustainable peace.

In essence, sustainable peace can be defined in terms of processes in a society that strengthen the attractor for positive interactions and keep social processes within its basin of attraction, while weakening the strength of the attractor for destructive conflict and violence.

Dynamical Strategies of Peace Making, Keeping, and Building

Our dynamical-systems approach has begun to generate a series of guidelines for use with promoting sustained peace on the ground. They include the following.

Simultaneous Management of War and Peace

As the attractor landscape model indicates, groups and communities typically hold the potential for dramatically different types of interaction patterns simultaneously. One attractor may capture the state of the system for extended periods of time (as is seen during protracted periods of conflict). However, this does not mean that peace building initiatives (peace education, dialogue groups, intergroup cultural exchanges, common community projects, etc.) during this period are for naught. The concept of latent attractors thus provides an important new perspective for understanding peace. In this view, the malignant thoughts, feelings, and actions characterizing a group's dynamics during conflict represent only the most salient and visible attractor for the group. Particularly, if there is a long history of interaction with the out-group, there may be other potential patterns of mental, affective, and behavioral engagement *vis a vis* members of the out-group, including those for positive relations.

With this in mind, identifying and reinforcing latent (positive) attractors, not simply disassembling the manifest (negative) attractors, should be the aim of conflict prevention in service of sustainable peace. In other words, in addition to attempts at achieving *negative peace* (an end to destructive conflict and violence), and the goal of *positive peace* (establishing fair systems of opportunity and justice), we must also strive to enhance *promotive peace*—the establishment of strong attractors for positive, constructive social relations. These objectives can be accomplished by implementing many of the initiatives summarized in the above multilevel framework.

Reverse Engineering Destructive Attractors

When conflicts do arise, the most obvious need is to quell any violence and actively contain destructive processes. This is often done by introducing police support, peacekeeping troops, or other forms of regional or international military interventions. However, even when systems de-escalate and appear to return to a state of peace, the potential for destructive interactions (destructive attractors) still exists. It is important, then, that we work actively to deconstruct and dismantle the negative attractors. In generic terms, the deconstruction of an attractor entails focusing on the elements comprising the pattern of behavior rather than focusing on the pattern itself. In the context of conflict, this means calling attention to specific actions, events, and pieces of information without noting their connection to the pattern in which they are embedded. When decoupled in this fashion, the lower-level elements may become re-configured into an entirely different pattern (e.g., a positive view of the outgroup and a benign or peaceful interaction pattern).

The important point is this: attacking the pattern itself is likely to intensify rather than weaken the pattern because of the tendency for attractors to resist change. One should focus instead on isolating elements and thereby weakening or eliminating the reinforcing feedback loops among them. As described in preceding chapters, there are several ways in which this can be accomplished in real-world settings, including: *introducing negative feedback loops (early-warning systems, cross-cutting structures, international monitoring, etc.); institutionalizing more nuanced, alternative conflict narratives (through media, textbooks, official accounts, etc.)*; and *limiting the pervasive spread of conflict by allowing movement of the parties.*

Increasing Complexity

Research has also shown that constructive social relations are characterized by relatively high levels of cognitive, emotional, behavioral, and structural complexity. Such complexity is advantageous when groups face problems or conflicts with other groups. As conflicts intensify, there is a strong tendency for the parties' thoughts, feelings, and behaviors to become more simple and black and white (which is evidence of strong attractor dynamics). If the conflict spreads to the community level and persists, then the same polarization occurs in social networks, groups, and institutions. However, communities and groups who maintain more complex cross-cutting (intergroup) structures and social networks, who hold more complex (multiple group) social identities, and who display more complex cognitive, emotional, and behavioral (adaptive) patterns, have been found to be more tolerant of outgroups, display less violence when conflicts spark, and engage in a more constructive manner when conflicts become difficult. Thus, *sustainable peace requires structures and processes that foster increased contact and complexity.*

Allowing Movement

The findings from our research support the basic idea that peace is associated with movement (Bartoli et al., 2010). When people and groups get trapped in narrow attractors for social relations, whether in patterns of destructive conflict, oppressor-oppressed dynamics, or even in patterns of isolation and disengagement from others, their well-being tends to deteriorate and their level of resentment tends to build. These traps may be constituted by physical structures such as segregated spaces, or by social-psychological constraints such as norms, attitudes, and ideologies. When trapped in such a well, people can become ever more destructive, oppressed, independent, and so forth, which acts to deepen the attractor and makes it less likely for them to escape its pull.

Of course, any pattern of behavior may be functional in certain situations; a destructive orientation fits very well in the discourses of armed conflict. But these patterns can become dominant and pervasive, so that when the current situation changes, or when people move to different situations, it is critical for people to adapt—to adopt different patterns of behaviors that are appropriate to the varied situations they face. From this perspective, *sustainable peace requires the establishment of conditions that allow for movement and adaptation.* At times, even "jiggling" the system through almost random movement, can break patterns and restore flexibility.

Every social system needs to change and evolve to remain functional; it needs to make constant adjustments to dissipate tensions. So if all the elements of the system (e.g., social groups) can have some influence on decisions, or at least opportunities for expression, the system evolves in small steps, enabling the release of tension. If, however, the capacity for movement (constant adjustments) is blocked, the repressed tension creates a strong negative attractor. At certain point, a sudden change occurs and the system dips into an attractor for destructiveness. All the factors that allow nonviolent expression and participation contribute to sustainable peace.

Movement Toward Justice

Decades of research on the psychology of justice have found that movement is also central to justice and peace. Relative deprivation, when people experience a gap in what they feel their group deserves and what it can achieve in comparison to similar groups, is central to conflicts over justice (Pruitt, 2006). This experience is typically triggered by change—shifts in the status quo that affects what groups expect, what they can get, and with whom they compare themselves. However, it is the need for *procedural justice* (Rawls, 1971), or the sense that there exist fair processes for the allocation of goods and for recourse against grievances, that has been shown to be critical to addressing injustice, even more so than actually receiving fair outcomes.

The *rate* at which justice is achieved is also critical. Peace scholars have found that minority groups who feel that the channels for fair recourse are blocked or unresponsive are more likely to revolt (Gurr, 2000). However, they have also found that when particular minority groups ascend to justice and equal treatment very rapidly, this can raise the aspirations, envy, and resistance of other groups (including those in power), and thus destabilize communities (Gurr, 2000; Lederach, 1997). Thus, *procedures of justice that provide a sufficiently steady response to the grievances of all stakeholders area necessary condition for sustainable peace.*

Developing Repellors for Violence

Anthropological research has shown that a key characteristic of peaceful societies, both historically and in the contemporary world, is the presence of non-violent values, norms, ideologies, and practices. Although non-violent norms are practiced in many communities around the globe, they are often overwhelmed by more violent ideologies, messages, and social modeling. There are a wide-variety of parenting and educational methods for fostering more non-violent, prosocial attitudes and skills in children, such as violence-prevention, tolerance, cooperative learning, conflict resolution, and peace education curriculum, just to name a few. However, *sustainable peace will require a much more concerted effort to teach non-violent values, norms, and practices to young people and to better limit exposure of youth to gratuitous forms of violence and to destructive social modeling by adults and public leaders.*

Acknowledging That Peace Is Never Achieved

Peace is a dynamic process that requires a set of fair processes and procedures that allow all stakeholders to negotiate for their needs and rights, in order to create unity out of diversity. Indeed, peace initiatives uninformed by an ongoing process of reading feedback and making necessary adjustments are destined to do more harm than good. Research has found that the most effective decision-makers are those who are able to continually adapt; by remaining open to feedback, they can reconsider their decisions and alter their course if necessary (Dörner, 1996). These leaders make more, not fewer, decisions as their plans unfold, and ultimately are able to enhance the well-being of the communities with which they work. Thus, effectiveness comes from flexibility not rigidity. In this way, *we can work to increase the probabilities that peace will emerge and be sustained.*

Conclusion: A Vision for Sustainable Peace

To conclude with the visionary dynamical words of John F. Kennedy:

> "Too many of us think [that peace] is impossible. Too many think it is unreal. But that is a dangerous, defeatist belief. It leads to the conclusion that war is inevitable, that mankind is doomed, that we are gripped by forces we cannot control. We need not accept that view. . .Let us focus instead on a more practical, more attainable peace, based not on a sudden revolution in human nature but on a gradual evolution in human institutions—on a series of concrete actions and effective agreements which are in the interest of all concerned. There is no single, simple key to this peace; no grand or magic formula to be adopted by one or two powers. Genuine peace must be the product of many nations, the sum of many acts. *It must be dynamic, not static, changing to meet the challenge of each new generation. For peace is a process—a way of solving problems.*"

References

Alban, C., & Druckman, D. (2012). The role of equality in negotiation and sustainable peace. In P. T. Coleman & M. Deutsch (Eds.), *Psychology's contributions to sustainable peace*. New York: Springer.

Barabasi, A. (2001). *Linked*. New York, NY: Penguin.

Bartoli, A. (2005). Learning from the Mozambique peace process: The role of the community of Sant'Egidio. In R. J. Fisher (Ed.), *Analyzing successful transfer effects in interactive conflict resolution*. Oxford: Lexington Books.
Bartoli, A., Bui-Wrzosinska, L., & Nowak, A. (2010). Peace is in movement: A dynamical systems perspective on the emergence of peace in Mozambique. *Peace and Conflict: Journal of Peace Psychology, 16*, 211–230.
Blum, G. (2007). *Islands of agreement: Managing enduring armed rivalries*. Cambridge, MA: Harvard University Press.
Coleman, P. T. (2003). Characteristics of protracted, intractable conflict: Toward the development of a metaframework—I. *Peace and Conflict: Journal of Peace Psychology, 9*, 1–37.
Coleman, P. T. (2012). The essence of peace? Toward a comprehensive and parsimonious model of sustainable peace. In P. T. Coleman & M. Deutsch (Eds.), *Psychological contributions to sustainable peace*. New York, NY: Springer.
Coleman, P. T., Bui-Wrzosinska, L., Vallacher, R. R., & Nowak, A. (2006). Protracted conflicts as dynamical systems. In A. K. Schneider & C. Honeyman (Eds.), *The negotiator's fieldbook: The desk reference for the experienced negotiator* (pp. 61–74). Chicago: American Bar Association Books.
Coleman, P. T., & Deutsch, M. (Eds.). (2012). *Psychological components of sustainable peace*. New York, NY: Springer.
Coleman, P. T., Hacking, A., Stover, M., Fisher-Yoshida, B., & Nowak, A. (2008). Reconstructing ripeness I: A study of constructive engagement in protracted social conflicts. *Conflict Resolution Quarterly, 26*, 3–42.
Coleman, P. T., Vallacher, R. R., Nowak, A., & Bui-Wrzosinska, L. (2007). Intractable conflict as an attractor: Presenting a model of conflict, escalation, and intractability. *American Behavioral Scientist, 50*, 1454–1475.
Coleman, P. T., Vallacher, R., Nowak, A., Bui-Wrzosinska, L., & Bartoli, A. (2009). A systemic approach to peace: Lessons from Mozambique. In M. Deutsch, P. T. Coleman, & E. Marcus (Eds.), *A guiding handbook for conflict resolution in the Arab world*. New York: International Center for Cooperation and Conflict Resolution Publications.
Coleman, P. T., Vallacher, R. R., Nowak, A., Bui-Wrzosinska, L., & Bartoli, A. (2011). Navigating the landscape of conflict: Applications of dynamical systems theory to protracted social conflict. In N. Ropers (Ed.), *Systemic thinking and conflict transformation*. Berlin, Germany: Berghof Foundation for Peace Support.
Deutsch, M. (1973). *The Resolution of Conflict: Constructive and destructive processes*. New Haven: Yale University Press.
Deutsch, M. (2006). A framework for thinking about oppression and its change. *Social Justice Research, 19*, 7–41.
Deutsch, M., Marcus, E., & Brazaitis, S. (2012). A Framework for thinking about developing a global identity. In P. T. Coleman & M. Deutsch (Eds.), *Psychology's contributions to sustainable peace*. New York: Springer.
Diehl, P., & Druckman, D. (2011). *Evaluating peace operations*. Boulder: Rienner.
Disney, A., & Gbowee, L. (2012). Gender and sustainable peace. In P. T. Coleman & M. Deutsch (Eds.), *Psychology's contributions to sustainable peace*. New York: Springer.
Dörner, D. (1996). *The logic of failure: Recognizing and avoiding error in complex situations*. Reading, Mass: Addison-Wesley Pub.
Fisher, R. E., Ury, W., & Patton, B. (1991). *Getting to yes: Negotiating agreement without giving in* (2nd ed.). New York: Penguin Books.
Fry, D. P. (2006). *The human potential for peace: An anthropological challenge to assumptions about war and violence*. New York, NY: Oxford University Press.
Fry, D. P., & Miklikowska, M. (2012). Culture of peace. In P. T. Coleman & M. Deutsch (Eds.), *Psychological contributions to sustainable peace*. New York, NY: Springer.

Gomes de Matos, F. (2012). LIF PLUS: The life-improving force of peaceful language use. In P. T. Coleman & M. Deutsch (Eds.), *Psychology's contributions to sustainable peace*. New York: Springer.

Gottman, J. M. (1993). The roles of conflict engagement, escalation, and avoidance in marital interaction: A longitudinal view of five types of couples. *Journal of Consulting and Clinical Psychology, 61*, 6–15. doi:10.1037//0022-006X.61.1.6.

Gottman, J. M., Swanson, C., & Murray, J. (1999). The mathematics of marital conflict. *Journal of Family Psychology, 13*, 3–19. doi:10.1037//0893-3200.13.1.3.

Gottman, J. M., Murray, C. D., Swanson, C. C., Tyson, R., & Swanson, K. R. (2002). *The mathematics of marriage*. Cambridge, MA: The MIT Press.

Gottman, J., Swanson, C., & Swanson, K. (2002). A general systems theory of marriage: Nonlinear difference equation modeling of marital interaction. *Personality and Social Psychology Review, 6*, 326–340.

Gratch, A. (2012). The psychodynamics of peace. In P. T. Coleman & M. Deutsch (Eds.), *Psychology's contributions to sustainable peace*. New York: Springer.

Gurr, T. R. (2000). *Peoples versus states*. Washington, DC: United States Institute of Peace Press.

Johnson, D. W., & Johnson, R. T. (1989). *Cooperation and competition: Theory and research*. Edina, MN: Interaction Book Company.

Johnson, D. W., & Johnson, R. T. (2005). New developments in social interdependence theory. *Psychology Monograph, 13*, 285–360.

Johnson, D., Johnson, R., & Tjosvold, D. (2012). Effective cooperation, the foundation of sustainable peace. In P. T. Coleman & M. Deutsch (Eds.), *Psychology's contributions to sustainable peace*. New York: Springer.

Koo, J., Han, J., & Kim, J. (2002). Integrative complexity of South–North Korean correspondences: A time-series analysis, 1984–1994. *Journal of Conflict Resolution, 46*, 286–304.

Koo, J., & Kim, J. (1999). Integrative complexity in newspaper columns and the bulletins of political parties: The impact of economic crisis and political transition on integrative complexity. *Korean Journal of Social & Personality Psychology, 13*, 35–52.

Kriesberg, L. (2005). Nature, dynamics, and phases of intractability. In C. A. Crocker, F. O. Hampson, & P. Aall (Eds.), *Grasping the nettle: Analyzing cases of intractable conflict* (pp. 65–98). Washington, DC: United States Institute of Peace.

Kugler, K. G., Coleman, P. T., & Fuchs, A. M. (2011). Conflict, Complexity, and Openness: Constructive versus Destructive Dynamics of Discussions over Intractable Issues. WOP Working Papers No. 2011/3. Retrieved from: www.psy.lmu.de/wirtschaftspsychologie/forschung/working_papers/index.html

Latane', B., & Nowak, A. (1994). Attitudes as catastrophes: From dimensions to categories with increasing involvement. In R. R. Vallacher & A. Nowak (Eds.), *Dynamical systems in social psychology* (pp. 219–249). San Diego: Academic Press.

Lederach, J. P. (1997). *Building peace: Sustainable reconciliation in divided societies*. Washington, DC: United States Institute of Peace Press.

Levine, R. A., & Campbell, D. T. (1972). *Ethnocentrism: Theories of conflict, ethnic attitudes, and group behavior*. New York: Wiley.

Liht, J., Suedfeld, P., & Krawczyk, A. (2005). Integrative complexity in face-to-face negotiations between the Chiapas guerrillas and the Mexican government. *Political Psychology, 26*, 543–552.

Lindner, E. (2012). Fostering global citizenship. In P. T. Coleman & M. Deutsch (Eds.), *Psychology's contributions to sustainable peace*. New York: Springer.

Losada, M. (1999). The complex dynamics of high performance teams. *Mathematical and Computer Modelling, 30*, 179–192.

Losada, M., & Heaphy, E. (2004). The role of positivity and connectivity in the performance of business teams: A nonlinear dynamics model. *American Behavioral Scientist, 47*, 740–765.

Mischel, W., DeSmit, A., & Kross, E. (2006). Self regulation in the service of conflict resolution. In M. Deutsch, P. T. Coleman, & E. Marcus (Eds.), *The handbook of conflict resolution: Theory and practice* (2nd ed.). San Francisco: Jossey-Bass.

Nets, R. (2012). Fixation and change of the Israeli official memory regarding the causes for the exodus of the Palestinian refugees during the 1948 War, 1949–2004 (Unpublished dissertation).

Nowak, A., & Vallacher, R. (1998). *Dynamical social psychology*. New York, NY: Guilford Press.

Nowak, A., Vallacher, R., Bui-Wrzosinska, L., & Coleman, P. T. (2006). Attracted to conflict: A dynamical perspective on malignant social relations. In A. Golec & K. Skarzynska (Eds.), *Understanding social change: Political psychology in Poland* (pp. 33–49). Haauppague, NY: Nova Science Publishers Ltd.

Pruitt, D. G. (2006). *A graphical interpretation of escalation and de-escalation presented at Dynamics and Complexity of Intractable Conflicts*. Poland: Kazimierz.

Rawls, J. (1971). *A theory of justice*. Cambridge: Harvard University Press.

Roccas, S., & Brewer, M. (2002). Social identity complexity. *Personality and Social Psychology Review, 6*, 88–106.

Sandy, S. V., & Boardman, S. K. (2000). The peaceful kids conflict resolution program. *International Journal of Conflict Management, 11*(4), 337–357.

Schwartz, I. B. (1992). Small amplitude, long period outbreaks in seasonally driven epidemics. *Journal of Mathematical Biology, 30*, 473–491.

Siegel, D. J. (2010). *Mindsight: The new science of personal transformation*. New York: Bantam.

Suedfeld, P., Tetlock, P. E., Streufert, S., Smith, C. P., Atkinson, J. W., McClelland, D. C., et al. (1992). Conceptual/integrative complexity. In C. P. Smith, J. W. Atkinson, D. C. McClelland, & J. Veroff (Eds.), *Motivation and personality: Handbook of thematic content analysis* (pp. 393–400). New York: Cambridge University Press.

Svyantek, D. J., & Brown, L. L. (2000). A complex-systems approach to organizations. *Current Directions in Psychological Science, 9*(2), 66–74.

Tetlock, P. E. (1985). Integrative complexity of American and Soviet foreign policy rhetoric: A time-series analysis. *Journal of Personality and Social Psychology, 49*, 1565–1585. doi:10.1037/0022-3514.49.6.1565.

Vallacher, R., Coleman, P. T., Nowak, A., & Bui-Wrzosinska, L. (2010). Rethinking intractable conflict: The perspective of dynamical systems. *American Psychologist, 65*(4), 262–278.

Vallacher, R. R., & Nowak, A. (Eds.). (1994a). *Dynamical systems in social psychology*. San Diego: Academic.

Vallacher, R. R., & Nowak, A. (1994b). The stream of social judgment. In R. R. Vallacher & A. Nowak (Eds.), *Dynamical systems in social psychology* (pp. 251–275). San Diego: Academic.

Varshnay, W. (2002). *Ethnic conflict and civic life*. New Haven, CT: Yale University.

Winter, D. G. (2007). The role of motivation, responsibility, and integrative complexity in crisis escalation: Comparative studies of war and peace crises. *Journal of Personality and Social Psychology, 92*, 920–937.

Chapter 8
Epilogue: Conflict in the Twenty First Century

Conflict is a defining feature of human relations. It would be naïve, if not irresponsible, to think that antagonistic interactions among individuals, groups, and nations can ever be eliminated or marginalized, no matter how much we learn about the root causes of such interactions. A realistic goal is to understand the fundamental dynamics that promote and sustain destructive conflict generally, and the special blend of dynamical properties that transform a small proportion of conflicts into a protracted state of hostile relations that destroys the fabric of interpersonal, intergroup, and international life. The book thus far has identified key dynamical principles underlying the progression toward intractable conflict, with an emphasis on conflicts observed throughout history and that are all too prevalent today. This understanding, in turn, provides the basis for divining means of disrupting the feedback loops that sustain protracted conflicts, so that such conflicts can be transformed into benign or even positive social relations.

Will such understanding and intervention strategies help mankind navigate the unchartered waters and likely challenges of the twenty first century? Conflict may well be a constant in human relations, but the world in which we live has experienced dramatic change over the course of human history, and the pace and quality of this change has accelerated in recent decades. Does the accelerating nature of social change have implications for the expression and intensity of conflict, or for mankind's ability to manage such conflicts before they pose insurmountable problems? On the one hand, there are good reasons to expect that destructive conflict will become an increasingly frequent and serious problem as we proceed through the twenty first century, and that such conflicts will prove increasingly difficult to contain and transform. But there are equally good reasons for anticipating the opposite: that conflict, although never eliminated, will become *less* frequent, destructive, and protracted as the dynamics of social life continue to evolve. Which of these opposing scenarios is likely to prevail on its own? To what extent can the dynamical perspective be employed to manage relations primed for conflict, so that the balance is tipped in favor of the optimistic scenario?

R.R. Vallacher et al., *Attracted to Conflict: Dynamic Foundations of Destructive Social Relations*, Peace Psychology Book Series, DOI 10.1007/978-3-642-35280-5_8,

The Case for Pessimism

When one thinks of the many proximate causes for destructive conflict, a rather dreary scenario can be envisioned. The most pressing cause of conflict is competition over scarce resources. For the majority of human history, this basis for conflict occurred when two neighboring tribes or villages both laid claim to the same grazing territory, the same irrigable land for pastures, or access to the same source of water. In today's world, such disputes are not confined to local groups but play out in the context of intra-national and international relations. The world's supply of fossil fuels, for example, is finite and could become effectively depleted in coming decades, yet the demand for this resource has never been greater and is likely to increase dramatically in the years to come as countries that were primarily agrarian and rural in the twentieth century transition to industrial and urban economies. There are alternatives to oil and natural gas (e.g., solar, hydrothermal, wind, nuclear), but these have yet to become developed in an economically—and politically—feasible manner. For the foreseeable future, then, a decreasing supply of fossil fuels coupled with increasing demand is a recipe for intense conflict within countries and for warfare between nations.

Fuel is important and necessary in today's world, where machines rather than human labor are the driving force in production and consumption. But in principle, one can find and exploit alternatives to fossil fuels. Water is an entirely different matter. No matter how technologically savvy and innovative we become, there is no alternative to water for sustaining human life. We are totally dependent on water for drinking, sanitation, irrigation, fishing, transport on oceans, lakes, and rivers, and power generation. Water, like fossil fuel, is a limited resource and the demands for it, driven by population growth, rising standards of living, and climate change, are increasingly dramatically. Indeed, the water crisis may emerge in the twenty first century as the primary basis for conflict over resources (Gleick, 1993).

Conflict is also fueled by the well-documented human proclivity to favor one's ingroup over outgroups (cf. Brewer, 1979; Tajfel, 1982). Especially when ingroup-outgroup relations center on ideology, norms, and customs, contact between groups is a breeding ground for stereotypes, negative feelings and attitudes, and violent confrontations. The world has always been diverse, with people in different regions showing marked differences on every conceivable dimension, from skin color and food preferences to economic systems and religion. Historically, however, physical distance and limited technology have prevented contact between people who differed on these dimensions and therefore did not unleash the potential for lethal inter-group contact.

The modern world has effectively erased the barriers between people in different parts of the world. One can reach virtually any spot on earth in less than a 12-h plane ride, and people are increasingly taking advantage of this possibility. Even without traveling, people are exposed to others of different races, religions, and cultures on a daily basis because of mass media (movies, television), the internet, and social media. Because of the self-selection of people to these forms of media, people are able to find and maintain social support for their views of outgroups—whether political, religious, or racial—and avoid exposure to information that might soften their views and promote tolerance (cf. Barabasi, 2001; Christakis & Fowler, 2009). The "echo chamber" effect of selective communication has the potential to enflame the competition and animosity inherent in ingroup-outgroup relations.

And then there is the issue of immigration. People in increasing numbers are moving to other countries and bringing their national identities with them. Rather than assimilating to the new culture, as was common in earlier times, immigrants today are increasingly inclined to embrace their values, religious traditions, languages, and customs in their adopted nation (Alba & Nee, 2005; Berry, 1997). Because immigrants and local residents live in a state of confined interdependence, the clash of cultures has the potential to tap the hostility latent in ingroup-outgroup relations.

Add to this mix the ongoing development of sophisticated weapons that can translate a simple conflict into an episode of destruction and death on an unimaginable scale. Fights between individuals and groups that were once fought with knives, clubs, spears, and swords are now fought with handguns and assault weapons. Interstate wars that were once engaged with troops and artillery are now fought with fighter planes, stealth bombers, ballistic missiles, and unmanned drones fired from a location thousands of miles away from the intended target. A battle that might produce a few thousand casualties can now decimate hundreds of thousands, if not millions of people. And the lack of face-to-face contact necessary to inflict harm on people reduces the constraints against the use of such means.

From a dynamical perspective, these factors can be viewed as elements that have positive feedback loops among them. The confined interdependence of different cultures amplifies competition over scarce resources, which in turn magnifies intergroup antagonism, which then reinforces the competitive approach to the resource dilemma. The availability of weapons of mass destruction in combination with their use at a safe distance increases the likelihood of destruction and death on scales unheard of in earlier times, which increases the desire of warring parties to stockpile and employ more deadly weapons. The increasing reliance on self-selected mass media and internet communication can reinforce outgroup stereotypes, which can in turn reinforce avoidance of information and contact that might undermine such stereotypes. These feedback loops do not seem to bode well for the frequency and intensity of conflict in the years to come.

The Case for Optimism

As plausible as this gloomy scenario seems, there is another way to view the frequency and intensity of conflict to be anticipated in the coming years and decades. Indeed, when violent conflict is viewed from a historical perspective, there is reason to be highly optimistic about the fate of interpersonal, intergroup, and international relations in the twenty first century. The simple fact is that violence of all forms has shown a dramatic decline over the past few centuries (Pinker, 2011). Tribal warfare was nine times as deadly as war and genocide were in the twentieth century. The homicide rate in medieval Europe was more than thirty times today's rate. Slavery, sadistic punishments, and executions were routine features of life for thousands of years, only to become targets for abolition in the last few centuries. Rape, battering, child abuse, animal cruelty, hate crimes, and deadly riots occur in today's world, but the frequency of each is dramatically lower today than in our past. The contemporary era, in fact, may be the least violent period in human history.

The decline in violence is attributable to a variety of factors that we can expect to become increasingly manifest as we move forward in the twenty first century. In his comprehensive analysis of the precipitous decline in interpersonal and intergroup violence, Pinker (2011) points to the rise and spread of central governments, increased intra-national and international trade, and the rise of urban centers and cosmopolitism. These factors have brought about self-control with respect to agreed-upon laws and norms, promoted bargaining and negotiation as a means of resolving disputes as alternatives to plundering and invasion, debunked toxic ideologies and superstitions, and encouraged the use of reason to reduce the temptations of violence and physical domination. The increased literacy occasioned by the rise of stable governments and enhanced standards of living has also contributed to the steady decline in violent confrontation because it has fostered empathy and perspective taking in social relations, and has familiarized people—at a safe distance—with the customs and traditions of different cultures.

One could argue, too, that the factors noted under the pessimistic scenario could work instead to make the world a *less* hostile and contentious place for interpersonal, inter-group, and international relations. The sharp increase in direct contact between people from different cultures in recent decades, for example, may serve to inhibit rather than activate the conflict inherent in ingroup-outgroup relations. When people fly to distant parts of the world, after all, they not only expect to observe different customs and norms, they are primed to appreciate these local features. Rather than competing with people in these cultures, visitors interact with them informally or through commercial transactions. The portrayal of individuals representing different races, ethnicities, and religions in mass media and movies, meanwhile, can also soften the tension associated with intergroup conflict because this contact is devoid of the conditions such as resource competition and confined interdependence that bring out the worst in ingroup-outgroup relations. One could argue, in fact, that when issues of resource scarcity do arise, they can serve as superordinate goals that serve to unify rather than divide people from different backgrounds around a common cause (cf. Sherif, Harvey, White, Hood, & Sherif, 1961). This assumes, of course, that the portrayal of diversity in entertainment and information media is largely positive and emphasizes variation rather than homogeneity within different groups—an assumption that seems to hold, at least in relatively affluent and democratic societies.

The contact between cultures may prove positive for yet other reasons. Recent decades have witnessed a sharp increase in emigration—movement from one country to settle in another country—and marriage between members of different ethnic groups and races. This has the effect of *increasing* the genetic and lifestyle variation within a culture and *reducing* these sources of variation between cultures. In the optimistic scenario, this blurring of group differences has the potential to reduce racism and ethnocentrism—which historically have been bases for intergroup prejudice and conflict.

The advances in technology witnessed in the past few years and decades may also contribute to an overall decline in destructive conflict. Such technology, of course, has given us weapons that can transform a heated dispute into a homicide or shooting rampage, not to mention weapons of mass destruction that can decimate entire populations. But for every technological device conducive to harm, there are many more that are conducive to benign and positive social relations. Almost everyone in the modern world has a smart phone or access to the internet, web-based communication, and social media. These technologies enable people to form connections on a worldwide basis, and to do so with limited risk of becoming involved in zero-sum relations that produce winners and losers. In effect, people are poised to become world citizens rather than xenophobic and nationalistic members of a tribe or country. This tendency may be enhanced by innovations in medical care, agriculture, education, and business technologies that can be shared across national borders and enhance collective well-being.

To be sure, people tend to self-select into virtual communities that echo their preconceived biases, beliefs, and concerns, and some bases for self-selection can serve to perpetuate stereotypes, misinformation, and intolerance. But people also use modern communication technology to build social networks that go well beyond the simple ingroup-outgroup dichotomies that dominated personal and social identity in the past. Today's world provides an enormous range of opportunities for self-expression, career choice, and other sources of identity creation. People may seek out others from different regions of the world because of a common hobby, artistic preference, economic agenda, or even because they wish to interact in a video game. These bases for social network formation not only displace shared identities rooted in tribalism and ethnicity, they also tend to be crosscutting in ways that prevent polarization between groups from developing (cf. Brewer & Brown, 1998; Roccas & Brewer, 2002; Varshnay, 2002). The person in another country with whom one discusses entertainment preferences may well have different family customs, political persuasion, or skin color.

There are important biological bases for violence and conflict, which would seem to ensure the continuation of malignant social relations despite the factors noted above. However, the connection between biology and conflict is mediated by other factors that can, in principle, weaken the connection. Historically, the proclivity for interpersonal aggression and intergroup violence has been a tendency launched and implemented largely by males (cf. Daly & Wilson, 1988, 1994). This may reflect the fact that males historically have had a monopoly on institutional power, so that females have not been in a position to make decisions regarding warfare, nor to implement such decisions. But another idea, rooted in biology, comparative psychology, and evolutionary psychology, is that males are inclined toward aggression because of the male sex hormone, testosterone. Research across many mammalian species has documented a connection between levels of testosterone and aggressive behavior (Ellis, 1986; Monaghan & Glickman, 1992). Testosterone levels presumably have been stable across human history, so perhaps we are doomed to violent conflict no matter how much the factors cited above come to the fore in the modern era.

Even here, however, there is cause for optimism. The immediate causal effect of testosterone is not aggression per se, but rather seeking and maintaining social dominance (e.g., Archer, 2006; Mazur & Booth, 1998). Aggression, of course, is a basic means of establishing dominance hierarchies. In infrahuman species, dominance entails threat displays, sexual mounting, and violent behavior intended to intimidate a rival. And throughout much of human history, individuals dominated one another by fighting, while groups asserted dominance by attacking one another with military forces. Among humans in the modern era, however, there are alternative means of establishing and maintaining dominance. Males—and females, for that matter—can acquire and display resources and symbols of power. They can rise to leadership positions in industry, science, and politics. Presumably, a person who has the opportunity to command respect in a government or a large corporation—or in sports or entertainment—can satisfy the testosterone-based need for dominance without physical confrontations and aggression. Nations, too, can achieve dominance through economic success, scientific and technological advances, and leadership in international relations, without resorting to military intervention.

Research relevant to this point has shown that the relationship between testosterone and aggression exists for men in lower socioeconomic classes whose opportunities for achieving dominance by other means is limited (e.g., Dabs & Morris, 1990). Among middle- and upper-middle class men, however, the correlation between testosterone and aggression vanishes. Presumably, men with opportunities for achievement and dominance in business, government, politics, and science can bypass the need to assert their strength through physical aggression.

In the coming decades, then, one might expect a decrease in the expression of male-based aggression as opportunities for success and prominence become more widely available within and between societies. Indeed, the same factor that historically has promoted horrific violence may turn out to be a driving factor in innovation and entrepreneurship as the new century unfolds. Of course, the expression of dominance could also take the form of totalitarian repression backed up by force. However, the forces noted earlier—democratic governments, cosmopolitism, literacy—are likely to provide a check on the rise of authoritarian regimes. The worldwide connectivity provided by smart phones, the internet, and social media also mitigate against the expression of dominance in this form.

Yet another factor with biological underpinnings is worth considering, though its likely impact on the reduction of violence in the modern era is considerably more speculative (Pinker, 2011). Violent behavior can be undertaken for instrumental reasons, but the most destructive forms of violence—homicide on an interpersonal level, genocide and slaughter on an intergroup level—represent aggressive impulses over which individuals have lost the executive mental control that is provided by conscious processes (Baumeister, Vohs, & Tice, 2007). Impulsive aggression that cannot be reigned in upon its instigation represents a breakdown in *self-control* (cf. Baumeister et al., 2007; Mischel, Shoda, & Rodriguez, 1989). Self-control—the inhibition of impulses, the delay of immediate gratification in order to achieve a delayed reward, the suppression of hostile feelings and unwanted thoughts—requires executive resources which override automatic and emotion-driven processes.

The deployment of executive resources, in turn, requires considerable mental energy, which is provided by a form of glucose called glycogen (Gailliot et al., 2007). Like the glucose that fuels muscle activity in the body, glycogen is a limited resource that is rapidly depleted by exercises of self-control. When such *ego depletion* occurs, a person's degree of self-control diminishes accordingly, and he or she has a difficult time inhibiting impulsive actions with an immediate payoff—despite their longer-term costs. A person who resists eating chocolate chip cookies in favor of radishes, for example, subsequently has a harder time persisting at difficult mental puzzles, or inhibiting the expression of hostile feelings towards someone who has irritated him or her (e.g., Baumeister, Bratslavsky, Muraven, & Tice, 1998; Tice, Bratslavsky, & Baumeister, 2001; Vohs, Baumeister, & Ciarocco, 2005; Vohs & Heatherton, 2000).

The capacity for self-control varies across individuals (Baumeister, Heatherton, & Tice, 1994; Mischel, Shoda, & Peake, 1988). But because it consumes metabolic resources, self-control may also vary across historical periods that differ in their respective availability of nutritional food sources. Food shortages have been the exception rather than the rule throughout history, and famines have been frequent and widespread in their occurrence. Perhaps the shortage of food historically has had the effect of reducing the glycogen necessary to fuel executive resources, thus putting people at risk for lapses in self-control under conditions of stress that call for restraint and rationality rather than impulse and emotionality (Vohs & Heatherton, 2000). If this reasoning has merit, one might expect the increasingly plentiful food supplies in the modern era—especially the ready availability of sugars and carbohydrates that are quickly converted into glucose in the body and brain—to fuel the self-control necessary to inhibit aggressive behavior that might otherwise be induced by competitive social relations, ingroup-outgroup relations, or clashes in values and ideology.

Which Scenario Will Prevail?

There is no guarantee that the trajectory of the last few hundred years, let alone the trajectory of recent decades, will continue in the same form in the coming years. In dynamical systems, what looks like a consistent pattern (or stable point) on one time-scale can prove to be entirely different when viewed over a longer time span. So the decline in violence observed since the middle ages, and seen more dramatically with the advent of international trade, increasing standards of living, and a greater base-rate of literacy, may not forecast what is in store for the twenty first century. Particularly in light of the accelerated rate of geophysical, social, and technological change in recent years, the decline in violence could show a dramatic (nonlinear) reversal as the modern era continues to unfold. Even when destructive conflict is not manifest with high frequency, it nonetheless exists as a latent attractor to which a social system can switch in response to various triggers.

The issue, then, is which scenario will prevail? Are the factors that hold potential for enhancing the likelihood of conflict destined to embroil nations and identity-based groups into conflict? Or do the factors that have minimized conflict and violence in recent decades and centuries promise to make the world an increasingly peaceful place?

The dynamical perspective offers some guidance—but no guarantee. To begin with, we can expect relatively stable patterns of social and international relations to emerge despite the rapid pace of social change and the enormous diversity of cultural, political, and economic agendas in today's world. Nonlinear dynamical systems are governed by a press for progressive integration of system elements into a higher-order state. Even in highly complex systems composed of many elements, there is a tendency for global states and patterns to emerge, with the elements linked by positive feedback loops. The emergent state or pattern functions as an attractor that provides for coherence and stability in the system, while resisting the potentially destabilizing forces of new elements. In the context of social relations, this means that despite the vast amount of relevant information and the continuous flow of events, over time the respective parties—whether individuals, groups, or nation-states—will establish a stable way of perceiving and acting toward one another.

The emergence of attractors, occasioned by the collapse of complexity, is enhanced under conditions of heightened importance, urgency, and stress (cf. Latane' & Nowak, 1994; Vallacher, Coleman, Nowak, & Bui-Wrzosinska, 2010). This suggests that under challenging or uncertain circumstances with high economic, political, or social stress, interpersonal and intergroup relations are especially predisposed toward attractor dynamics, with each party viewing and responding to one another in evaluatively coherent terms (e.g., good vs. bad, friend vs. enemy). Because of the feedback among the many factors comprising a social system, the activation of a single factor (e.g., an economic downturn, resource competition) can activate all other factors (e.g., ideological differences) and produce a disproportionate response (e.g., a dramatic swing in public opinion) to minor events (e.g., a news report, an unsubstantiated rumor) that might otherwise have limited impact. Because challenges and stresses are likely to be prevalent in the twenty first century, we can expect the collapse of complexity and formation of strong attractors for social relations.

Another hallmark of nonlinear dynamical systems, however, is their potential for bifurcation—the same set of factors can result in the emergence of qualitatively different higher-order states depending on the dynamics of the system in question (cf. Kelso, 1995; Nowak & Vallacher, 1998; Ruelle, 1989). Thus, whether the proximate factors historically associated with conflict and the factors that have assumed prominence in recent history combine to enhance versus diminish the frequency and intensity of destructive conflict depends on how these factors are linked dynamically by means of their mutual influence. One can envision the emergence of positive social relations (in line with the optimistic scenario) in response to a new stressor, but one can just as reasonably expect the emergence of malignant social relations (in line with the pessimistic scenario). For that matter, one can envision some pattern of temporal oscillation—periodic, quasi-periodic, or chaotic—between these very different scenarios as the twenty first century unfolds.

Which scenario unfolds may not be entirely beyond prediction and control, however. To be sure, coherent higher-order states commonly emerge by means of self-organization, lending a noteworthy degree of unpredictability to the outcome. Lacking outside guidance, seemingly minor events can be determinative, shifting the balance in one direction or the other because of the amplifying effects of positive feedback. But emergence can also be guided by external forces and information—although only if these sources of influence are provided when the system is a state of disassembly (Vallacher & Wegner, 2012) and thus "ripe" for transformation (Coleman, 2000; Pruitt, 1997; Zartman, 2000). The challenge for the twenty first century is to provide the right cues for meaning at the right time. In the hands of malevolent forces that play to people's prejudices and fears, the future could relaunch patterns of entrenched conflict and destructive violence. But in the hands of others with the long-term benefits of all in mind, the optimistic scenario could play out. Now, perhaps more than ever, the fate of mankind is in our hands.

References

Alba, R. D., & Nee, V. (2005). *Remaking the American mainstream: Assimilation and contemporary immigration*. Cambridge, MA: Harvard University Press.

Archer, J. (2006). Testosterone and human aggression: An evaluation of the challenge hypothesis. *Neuroscience and Biobehavioral Reviews, 30*, 319–345.

Barabasi, A. (2001). *Linked*. New York, NY: Penguin.

Baumeister, R. F., Bratslavsky, E., Muraven, M., & Tice, D. M. (1998). Ego depletion: Is the active self a limited resource? *Journal of Personality and Social Psychology, 74*, 1252–1265.

Baumeister, R. F., Heatherton, T. F., & Tice, D. M. (1994). *Losing control: How and why people fail at self-regulation*. San Diego: Academic.

Baumeister, R. F., Vohs, K. D., & Tice, D. M. (2007). The strength model of self-control. *Current Directions in Psychological Science, 16*, 351–355.

Berry, J. S. (1997). Immigration, acculturation, and adaptation. *Applied Psychology, 46*, 5–34.

Brewer, M. B. (1979). In-group bias in the minimal intergroup situation: A cognitive-motivational analysis. *Psychological Bulletin, 86*, 307–324.

Brewer, M. B., & Brown, R. J. (1998). Intergroup relations. In D. T. Gilbert, S. T. Fiske, & G. Lindzey (Eds.), *The handbook of social psychology* (4th ed., Vol. 2, pp. 554–594). New York, NY: McGraw-Hill.

Christakis, N. A., & Fowler, J. H. (2009). *Connected: The surprising power of our social networks and how they shape our lives*. New York, NY: Little, Brown and Company.

Coleman, P. T. (2000). Fostering ripeness in seemingly intractable conflict: An experimental study. *International Journal of Conflict Management, 11*, 300–317.

Dabs, J. M., Jr., & Morris, R. (1990). Testosterone, social class, and antisocial behavior in a sample of 4462 men. *Psychological Science, 1*, 209–211.

Daly, M., & Wilson, M. (1988). *Homicide*. New York, NY: Aldine deGruyter.

Daly, M., & Wilson, M. (1994). Evolutionary psychology of male violence. In J. Archer (Ed.), *Male violence* (pp. 253–288). New York, NY: Routledge.

Ellis, L. (1986). Evidence of neuroandrogenic etiology of sex roles form a combined analysis of human, non-human primate and nonprimate mammalian studies. *Personality and Individual Differences, 7*, 519–552.

Gailliot, M. T., Baumeister, R. F., DeWall, C. N., Maner, J. K., Plant, E. A., Tice, D. M., et al. (2007). Self-control relies on glucose as a limited energy source: Willpower is more than a metaphor. *Journal of Personality and Social Psychology, 92*, 325–336.

Gleick, P. H. (1993). Water and conflict: Fresh water resources and international security. *International Security, 18*, 79–112.

Kelso, J. A. S. (1995). *Dynamic patterns: The self-organization of brain and behavior*. Cambridge, MA: MIT Press.

Latane', B., & Nowak, A. (1994). Attitudes as catastrophes: from dimensions to categories with increasing involvement. In R. R. Vallacher & A. Nowak (Eds.), *Dynamical systems in social psychology* (pp. 219–249). San Diego: Academic.

Mazur, A., & Booth, A. (1998). Testosterone and dominance in men [target article and commentaries]. *The Behavioral and Brain Sciences, 21*, 353–397.

Mischel, W., Shoda, Y., & Peake, P. K. (1988). The nature of adolescent competencies predicted by preschool delay of gratification. *Journal of Personality and Social Psychology, 54*, 687–696.

Mischel, W., Shoda, Y., & Rodriguez, M. L. (1989). Delay of gratification in children. *Science, 244*, 933–938.

Monaghan, E., & Glickman, S. (1992). Hormones and aggressiveness behavior. In J. Becker, S. Breedlove, & D. Crews (Eds.), *Behavioral endocrinology* (pp. 261–285). Cambridge, MA: MIT Press.

Nowak, A., & Vallacher, R. R. (1998). *Dynamical social psychology*. New York, NY: Guilford.

Pinker, S. (2011). *The better angels of our nature: Why violence has declined*. New York, NY: Penguin.

Pruitt, D. G. (1997). Ripeness theory and the Oslo talks. *International Negotiation, 2*(2), 237–250.

Roccas, S., & Brewer, M. B. (2002). Social identity complexity. *Personality and Social Psychology Review, 6*, 88–106.

Ruelle, D. (1989). *Elements of differentiable dynamics and bifurcation theory*. New York, NY: Academic.

Sherif, M., Harvey, O. J., White, J., Hood, W., & Sherif, C. W. (1961). *Intergroup conflict and cooperation: The robber's cave experiment*. Norman: Institute of Intergroup Relations, University of Oklahoma.

Tajfel, H. (1982). *Social identity and intergroup relations*. Cambridge, England: Cambridge University Press.

Tice, D. M., Bratslavsky, E., & Baumeister, R. F. (2001). Emotional distress regulation takes precedence over impulse control: If you feel bad, do it! *Journal of Personality and Social Psychology, 80*, 53–67.

Vallacher, R. R., Coleman, P. T., Nowak, A., & Bui-Wrzosinska, L. (2010). Rethinking intractable conflict: The perspective of dynamical systems. *American Psychologist, 65*, 262–278.

Vallacher, R. R., & Wegner, D. M. (2012). Action identification theory. In P. A. M. Van Lange, A. W. Kruglanski, & E. T. Higgins (Eds.), *Handbook of theories of social psychology* (pp. 327–348). Thousand Oaks, CA: Sage.

Varshnay, W. (2002). *Ethnic conflict and civic life*. New Haven, CT: Yale University Press.

Vohs, K. D., Baumeister, R. F., & Ciarocco, N. (2005). Self-regulation and self-presentation: Regulatory resource depletion impairs impression management and effortful self-presentation depletes regulatory resources. *Journal of Personality and Social Psychology, 88*, 632–657.

Vohs, K. D., & Heatherton, T. F. (2000). Self-regulatory failure: A resource-depletion approach. *Psychological Science, 11*, 249–254.

Zartman, I. W. (2000). Ripeness: The hurting stalemate and beyond. In P. C. Stern & D. Druckman (Eds.), *International conflict resolution after the Cold War* (pp. 225–250). Washington, DC: The National Academy of Sciences.

Appendix A: Design for Workshops on the Application of Dynamical Systems to Intractable Conflict

Change agents cannot bring intractable conflicts to an end. But they can do a great deal to help stakeholders broaden their sense of the forces driving and constraining the conflict, identify opportunities for destabilizing long-term patterns, locate and support existing networks of effective action, and help both reduce the probabilities that conflicts will intensify and persist and increase the probabilities that they eventually settle into more constructive patterns. Therefore, third parties and stakeholders need to learn how to think differently about the ways such evolving systems can be changed.

What follows is *not* the only way to conduct a DST applications workshop; the reader should consider this an additional collaborative voice that can inform their own work on design and implementation. And, of course, given that human beings will be involved, they should also stay open to changing and adjusting any design as needed.

Pre-session Participant Preparation. Given the relatively new and unfamiliar nature of the DST approach to conflict, we suggest sending out a pre-session message to each of the participants which helps frame expectations and provides some conceptual background for the approach. We recommend that it: (1) communicate the seriousness with which you are approaching the case and their concerns, (2) manage expectations by framing the sessions as unique—unlike other mediation, dialogue, etc. experiences they may have experienced in the past, and (3) provide them with user-friendly content about the general approach you plan to take. For example, you could instruct them to view "short videos about DST, session overviews, and managing complexity.[1] You might also include links to blog posts,[2] depending on the type of stakeholders involved." We also recommend

[1] For a series of videos by Dr. Peter Coleman on intractability visit: http://www.youtube.com/ICCCRTC. For a video of a DST session overview visit: http://www.youtube.com/watch?v=c7PrLXSIt3o. An additional video on managing complexity can be accessed here: http://www.ted.com/talks/lang/en/eric_berlow_how_complexity_leads_to_simplicity.html.

[2] Some useful blog posts include: "The Mathematics of Middle East Conflict and Peace" by Peter T. Coleman (http://www.psychologytoday.com/blog/the-five-percent/201110/the-mathematics-middle-east-conflict-and-peace), and "New ways to think about solving intractable problems" by

R.R. Vallacher et al., *Attracted to Conflict: Dynamic Foundations of Destructive Social Relations*, Peace Psychology Book Series, DOI 10.1007/978-3-642-35280-5,

contacting each participant individually before the session to see if they have had time to review this material and if they have any questions. It will help demystify the approach and communicate a sense of importance and care.

Phase 1: Framing and Overview of Session. One facilitator should begin the session by introducing the members of the team and providing an overview of the work—referring back to the prep-information on DST that was previously sent and emphasizing the importance of the fact that this experience may be different from other previous ones. Stress that this approach is NOT about solving the problem, mending fences or simply feeling better—but about understanding the dynamics of the conflict they are stuck in as accurately as possible and guiding longer-term change. The facilitators should walk the participants through a sort of agenda, providing a loose sense of the proposed sequence of events. They should also invite the participants to introduce themselves, to express their session and long-term objectives (you might list these on a flip chart). Above all, the speaker needs to communicate professionalism, care, openness and respect for their concerns.

Phase 2: Complicate. The objectives of phase 2 are to allow the stakeholders to tell their narratives within their small constituent groups—framing, analyzing and visualizing sequence of events that got them here through the use of *conflict-mapping*. Facilitators should walk the participants through a simple illustration of how to map a series of events, how to connect them with arrows, links and loops that show how each event made subsequent events better or worse over time. If possible, the participants should be provided with large flip charts, color markers or similar resources for this (however, some of our colleagues have drawn such maps in sand and dirt while working out in the bush, using twigs and stones to represent elements and relationships). This is a critical stage, so provide enough time for participants to generate a sufficient first draft.

The basic steps for mapping include:

- **Step 1**: Begin with the stakeholders, clarify which dynamic they are trying to understand (stalemate, escalation, latent opportunities for peace, events that might happen in the future, etc.).
- **Step 2**: Next, guide the stakeholders as they identify the chronology of events relevant to the dynamic of interest as far back as they feel is important, leading up to the present circumstances.
- **Step 3**: Then, instruct the stakeholders begin to map and connect the different events—in chronological order (see Coleman, 2011, pp. 117). For a map sample, refer to Fig. 1.

Sarah Lutman (http://www.artsjournal.com/speaker/2011/05/new-ways-to-think-about-solving-intractable-problems/).

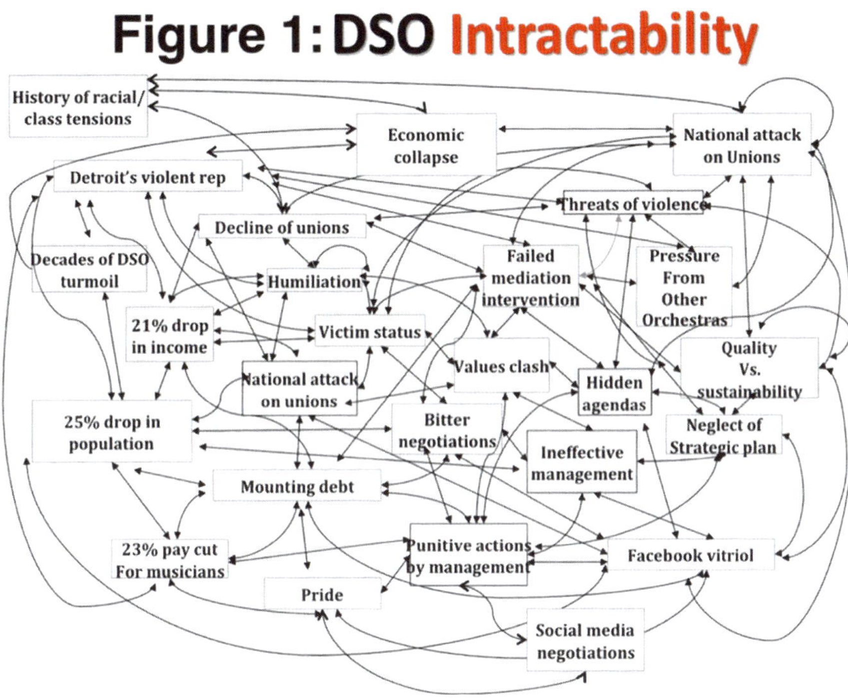

- **Step 4**: If mapping *conflict* dynamics (as opposed to peace potential), make sure to ask participants, the following questions: "What keeps the conflict from getting worse? What elements contain the conflict? What groups, actions or events help and keep it from getting worse?" Have them add these to the map. Also: "Thinking back, were there any major incidents—social or political shocks—that occurred 2, 5, even 10 years back that may have played a role in these dynamics?" Map this. "How have previous attempts to solve this played a role?" Map. These maps may look like the following, which include mitigating elements (islands of agreement) that help to contain the conflict:

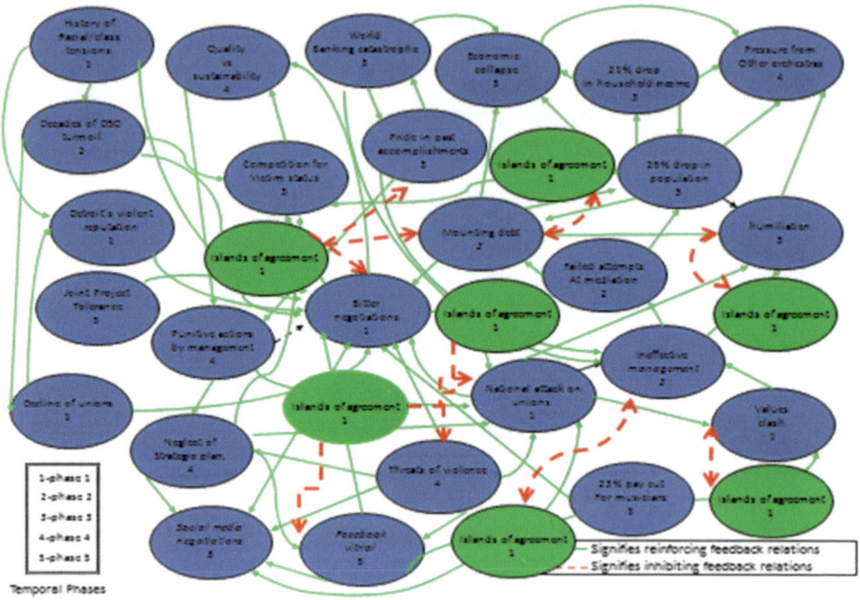

- **Step 5**: When they have completed the preliminary maps, ask the participants to stand back and take a few moments in their groups to view their map and see if they notice anything surprising or new. There may be differences of opinion or conflict within the groups, which is good to get at now. Help them work through this in the small groups.
- **Step 6**: Next, ask if they would be willing to walk around the room and view the maps generated by the different stakeholder groups. Suggest they do this silently and take notes about their thoughts and feelings as they walk around. Then bring them back together as a large group and debrief what they saw and learned. Allow this to go on as long as it is constructive. Then instruct them to return to their original groups and maps and add or change anything they wish in their original maps.

Note however that when employing a conflict-mapping methodology it is important that participants are clear about the degree of *objectivity* versus *subjectivity* of the exercise, as well as of the elements that are being mapped. Conflict maps can be generated based on actual, empirically verifiable entities, structures, processes, and events (such as communications networks) or on disputant's and other stakeholder's *perceptions* of relevant parties, issues, norms, events, etc. in a conflict. This latter form is known as *concept-mapping* or *mind-mapping* and is an excellent exercise for unearthing assumptions, perceptions and misperceptions. It may also provide a sense of context and nuance to the perceiver's sense of a conflict and allow for the exploration of temporal dynamics in people's understanding of the chronology of events—but it should not be mistaken for objective facts or truths.

Phase 3: Simplify. Next, using one of the maps—begin to help the participants learn to identify aspects of their maps that help to simplify and focus their understanding of the system. For example: identifying central *hubs* of activity, elements that link with many other elements; key *reinforcing loops*, elements that stimulate themselves through links with other elements in an ongoing fashion; and the *ratio,* or balance, between conflict-reinforcing and -inhibiting feedback, which can determine whether a conflict is escalating, deescalating, or stuck in a stalemate. Also, the identification of *actionable hubs* can be useful, more local elements and links that can feasibly be addressed. Identifying hubs and feedback loops in the system is particularly important for locating *centers of energy in the system, gateways for high-impact intervention, strategic targets for introducing conflict-inhibiting feedback* (such as early-warning systems that deter escalation), and *peace-reinforcing feedback* (like high-stakes common interests that motivate reconciliation). They can also help focus the analysis of conflict-mapping and manage the anxiety associated with the overwhelming sense of complexity of the system. However, they can do so in a manner informed by its complexity.

Working in the original small groups, the participants should begin to identify areas that they feel it might be particularly useful to focus their thinking and discussion on (refer participants to the video on complexity that was mentioned earlier). Once they feel they have identified 1-2 actionable areas they wish to discuss, bring them back together as a large group and have them share their thinking. Ideally, you use this information to begin to identify areas they can begin to work on in Phase 4.

Phase 4: Build Up. Ideally, here you gather together all stakeholders in mixed-stakeholder groups (small groups, even if all stakeholders are not represented in each group—just make sure the major divisions in the large conflict are together in each of the small groups). Then, ask them to begin to envision—5, 10 years out—the components—processes, institutions, etc.—that they consider should be in place need to be in place to allow stakeholders to begin to live and work together more constructively. If possible, they should focus the discussions on the areas identified in the previous phase—perhaps one for each group. How do they build a more just, safe, fair and healthy future together? You may need to provide examples and illustrations of what others have done to address the areas they are focusing on, such as:

- *Support latent networks of effective action.* Almost every conflict system will have people and groups who might be able to reach out across the divides and work to foster dialogue and peace but are constrained by the dynamics of the conflict. During times of intense escalation, these people and groups may become temporarily inactive—even going underground (a component of latent attractors)—but are often willing to re-emerge when conditions allow, becoming fundamental players in the transformation of the system.
- *Mobilize weak power facilitators.* Strong conflict systems are typically comprised of stable states of hostility, strong attractors for destructiveness and weak attractors for peace, and will typically reject most attempts to promote

peace. Nevertheless, change can emerge at the margins through non-threatening individuals, groups and communication processes that allow some key actors in the system to begin to consider alternatives to the status quo. This initial consideration is often made possible by the 'weaknesses' of certain propositions and proponents.

- *Employ negotiation chains.* A popular tactic used to initiate peace talks in protracted conflicts is the practice of involving a sequence of actors in the exploration of more formal talks, which allows each actor to speak directly with another actor with whom they are not constrained politically against speaking, but who has contacts further down the chain with the other side. Thus, talks transpire through a series of encounters, which allow for communications between parties who (1) need to be able to maintain deniability in the talks, and (2) who would otherwise not be able to communicate. This tactic allows for movement and communication in systems that are otherwise tightly controlled and constrained.
- *Work on positivity away from conflict attractors.* Recognizing that systems with strong intractable conflict attractors often construct peace-makers as part of the conflict system and position them in one camp or another, some interveners attempt to work constructively by circumventing the conflict (Praszkier, Nowak, & Coleman, 2010). This tactic aims to reduce the misery associated with these situations, but does so in a manner that is framed as external to the conflict. Some development efforts achieve this.
- *Acknowledge superordinate identities and goals.* This is a classic approach to addressing intergroup conflict that involves the identification or development of joint goals and identities in an attempt to establish a foundation of cooperation and eventually trust between parties (Sherif, Harvey, White, Hood, & Sherif, 1961; Deutsch, 1973; Worchel, 1986). Even if peacekeeping missions, reconciliation processes, trust-building activities, and cooperative conflict resolution initiatives appear to be largely ineffective in situations locked in an ongoing protracted struggle, they may very well be acting to establish a sufficiently wide and deep attractor basin for moral and humane forms of intergroup interactions that provide the foundation for a stable more peaceful future. The gradual and long-term construction of a positive attractor may be imperceptible, but it prepares the ground for a positive state that would be impossible without these actions.

If there is time available, debrief as a large group.

Phase 5: Tear Down. Now, ask the participants to imagine that they have arrived at this better place in 10 years. Ask them to identify those things that could easily suck them back into destructive conflict and violence. What are the traps that lay in wait? And what are the mechanisms and policies and institutions that would need to be in place to help keep these things from happening. Again, an illustration or two around early warning systems, regulations, more balanced narratives, etc., will help. Some examples include the following:

- *Decouple conflict-reinforcing feedback loops*. Depending on the nature of the conflict, disassembling the structure of the conflict may take different forms. If the structure of conflict binds together perceptions of all out-group members, showing positive examples of specific out-group members can increase complexity—since a single judgment cannot accommodate all the outgroup members. Another tack is to find an important (e.g., high status, charismatic) in-group member who doesn't share the ingroup's view of the conflict. If this person is sufficiently central that he or she cannot be marginalized within the group, the homogeneity of the in-group's perspective will be destabilized.
- *Introduce conflict-inhibiting feedback loops (early-warning systems, cross-cutting structures, international monitoring*, etc.). Once the positive feedback system of conflict escalation is mapped, it can help target specific links for the introduction of negative feedback mechanisms.
- *Institutionalize more nuanced, alternative conflict narratives (through media, textbooks, official accounts*, etc.). Strong war systems typically result in distinct and polarized narratives about the history of the conflict: who played the roles of heroes and villains, and what is still at stake. Mechanisms to monitor and revise such one-sided narratives are important for preventing future generations from returning to the same destructive patterns.

Phase 6: Change. Now what? What should the participants do first? How should they start? This phase involves a combination of teaching and problem-solving with each group or the large group, as they start to contemplate how to begin to implement their thinking moving forward.

For example, informing them that one recurring way that instability occurs in protracted conflicts is through major shocks to the system. Research on international conflict has found that significant political shocks are associated with both the onset and the resolution of enduring conflicts (Diehl & Goertz, 2001). The shock may be a world or civil war, a significant change in territory and power relations, regime change, an independence movement, or a transition to democracy. In the personal realm, this translates to a loss of employment, weather-related disaster, or a family crisis.

There are a series of tactics related to understanding and triggering change (see Coleman, 2011, pp. 171) that, depending on where the participants are in the process and what they are discussing, may allow them to know and think about ways to move forward. The facilitation of this subgroup may need to be the most flexible and adaptive in terms of the tactics employed—but remember, the objective is to help the group as a whole in thinking about the *what now* and *what next*.

Phase 7: Stabilize. This final phase essentially involves working with the stakeholders to help them understand the cautionary tale about the unintended consequences of even well-intended actions, common planning, and decision-making traps, as well as providing them with recommendations for more effective decision-making and adaptation within complex systems.

In the book *The Logic of Failure: Why Things Go Wrong and What We Can Do to Make Them Right* (1996), Dietrich Dörner presents his research on decision making and initiating change in complex environments using "simulated communities." A standard dynamic found in his research is that existing short-term problems are solved without giving sufficient thought to repercussions or new problems the solutions might create. This research offers evidence of more effective methods of decision making and problem–solving with complex systems, such as:

- **Making more decisions**: Effective decision-makers tended to assess a situation and set a course, but then continually *adapted*, staying open to feedback and to reconsidering their decisions and altering their course. They were found to make more, not fewer, decisions as their plans unfolded. They found more possibilities for enhancing their community's well-being as the situation evolved.
- **Acting more complexly**: These leaders seemed to understand that the problems they were addressing were closely linked with other problems, and that so their actions can have multiple effects. Therefore, they made many more decisions and actions when attempting to achieve one goal. In contrast, those who failed would typically make one decision per goal.
- **Focusing on the real problems first**: These decision-makers took the time to gather enough information to determine the central problems to address, and did not jump into action prematurely or simply focus on the problems they *could* solve and wanted to because it felt good to feel competent.
- **Testing hypotheses more**: Effective leaders also tested their solutions in pilot projects and assessed the effects before committing to them.
- **Asking why? more**: They also actively investigated the *why* behind events: the causal links that constitute the *network of causation* in their community.
- **Staying focused on the prize**: Ineffective decision makers got easily distracted and diverted; they hopped around a lot from problem to problem as each arose. Effective decision makers identified the central issues early on and stayed focused on addressing them.
- **But not on one solution**: However, effective decision makers did not develop a preoccupation with one solution; if feedback informed them that a solution was too costly or ineffective, they altered their approach.

Dörner's (1996) approach to problem solving is open, flexible, and tolerant of ambiguity. It requires reflection on *how* we are thinking and solving problems, as well as a keen recognition that data matters. These ideas can be introduced to stakeholders as parables (for example, by reading Dorner's Chernobyl case), or by simply by describing how-tos regarding problem-solving, feedback, and adaptation.

Phase 8: Closing and Next Steps. Ideally, you want to make sure that the stakeholders are aware of their accomplishments, how far they went, and what lies ahead. It is also important for there to be a clear sense of what's next and how your team can stay involved to support them moving forward. Some type of large group debrief might work, depending on the specifics of the situation. Remember, although this involves both art and science, at this point you are more deeply immersed in artistry.

Appendix B: Simulation of Attractor Dynamics

Based on the three scenarios by which an attractor landscape can be transformed (Chap. 6), we have developed a simulation platform to be employed when negotiating the resolution of difficult conflicts (Nowak et al., 2010a). We first describe the simulation program and its user interface, then describe how it can be utilized as a teaching tool and as platform for achieving a shared reality regarding conflict in negotiations. We also present preliminary evidence regarding its added value compared to traditional methods of negotiation.

Simulation of Attractor Dynamics

The development of the attractor simulation was inspired by our experiences working with schools experiencing chronic conflict and violence. We would typically begin these initiatives by interviewing stakeholders in schools to generate a sense of the problems they faced. These interviews often elicit a host of interrelated problems and processes, leaving the stakeholders and our team feeling overwhelmed by the complexity and intractability of the problem. The multitude of factors influencing the situation, and the multitude of consequences of each factor on the whole system's dynamics made any clear analysis of these conflict situations or the development of plausible scenarios for successful intervention seemingly impossible. Almost every possible intervention targeted one factor or another in the conflict, but failed to address other issues, and could not be undertaken given the conditions of the whole system.

The main goal of the attractor software was to help conflict stakeholders, negotiators, and third parties understand and systematically map the complexity of factors influencing such conflict systems, in order to visualize possible intervention strategies, with a consideration of the dynamical consequences and potential impact of their actions.

R.R. Vallacher et al., *Attracted to Conflict: Dynamic Foundations of Destructive Social Relations*, Peace Psychology Book Series, DOI 10.1007/978-3-642-35280-5,

The Attractor Software

The attractor software is a visualization tool. It prompts the user to specify key factors relevant to conflict case and actions that can be undertaken to influence the conflict. The user is then promoted to estimate the influence of each factor or action with respect to three considerations. Specifically, the user estimates how much the factor or action would influence (1) *the current state of the conflict* in a positive or negative direction, (2) *the potential for future conflict or negative interactions*, and (3) the *potential for positive social interactions*. In other words, the user is asked to predict the immediate impact of an action and both the negative and positive long-term consequences of the action (see the interface, below). The software merely visualizes the user's understanding of these consequences, based on his or her expertise and experience with the case in question. It is up to the user—or a group of users or stakeholders engaged in negotiation—to specify the case and the relations to be analyzed (e.g., a marriage, an ethnic conflict).

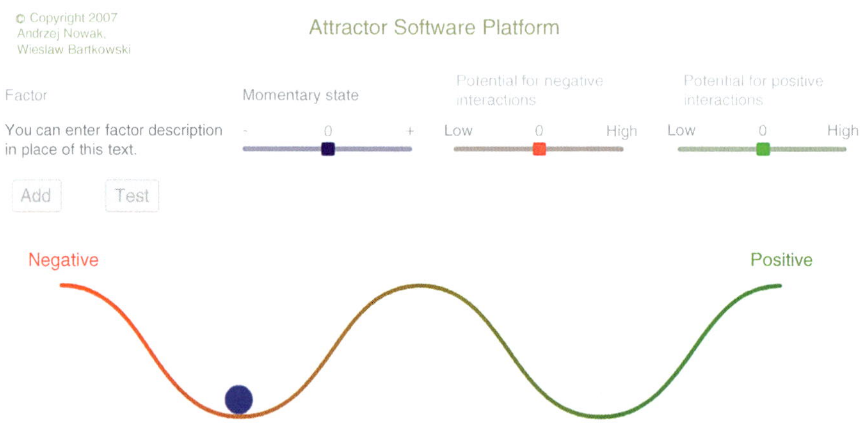

Interface for the Attractor Software

The program depicts two attractors for the conflict: a positive attractor indicating stabilization of favorable attitudes and positive actions, and a negative attractor indicating stabilization of unfavorable attitudes and negative or violent actions. If the current state of the relationship is within the basin of a strong attractor, this state is unlikely to change despite the introduction of factors relevant to change. Conversely, if the current state of the relationship is outside the basin of attraction, the relationship may display a qualitative change (e.g., from positive to negative) with the addition of a single, seemingly unimportant factor.

Using the Software

The users type in the label for each factor or action and use a slider bar to specify its importance in affecting the relationships within the system. It is assumed that the user is knowledgeable about the nature and importance of factors that are relevant to the relationship. Someone who works with high school gangs, for example, is in a position to identify the various conditions and triggers that affect each gang's actions and the inter-gang relationship. Despite the user's expertise and insight, however, the influence of the specific factors introduced into the software is not obvious. As noted earlier, a minor provocation can push two groups into open warfare, whereas a major change in conditions might have little effect on the relations between them. Such non-linearity between influencing factors and observable consequences follow from the nature of attractor dynamics. If the current state of a relationship is captured by an attractor, even strong external forces aimed at changing the status quo may be countered by the attracting tendency of prevailing norms and feelings. But if the current state is positioned outside the basin of one attractor, even a minor force might be sufficient to move the relationship toward a different attractor.

The user then estimates the likely impact of each factor with respect to the considerations noted above (immediate, long-term negative, and long-term positive). Estimating positive and negative consequences separately is critical because a dynamical system may be characterized by two conflicting attractors (e.g., one negative, one positive). In a marriage, for example, raising children can strengthen the bonds between parents, but it can also produce stress and challenge their relationship. It is important to note that factors relevant to a conflict can operate in accordance with very different time scales. In ethnic relations, for example, income disparity is a very important factor in the long run, but it is unlikely to spark an episode of violence on a particular day. Asking the user to estimate both immediate and long-term consequences is thus an important feature of this method.

Workshops Employing the Software

The attractor software allows users to integrate and solidify their understanding of conflict from a dynamical systems perspective. It also can provide practical experience in dealing with real cases of complex conflicts. We have prepared an accessible tutorial designed to introduce the notions of dynamical systems, attractors, and conflict dynamics to an audience unfamiliar with this perspective (see http://www.dynamicsofconflict.iccc.edu.pl). We have also employed the attractor software in workshops and classes (e.g., Columbia University, West Point Military Academy) to introduce the dynamical systems approach to conflict.

In these workshops and classes, we begin by presenting the software interface and linking the visualization of attractors to relevant features of dynamical systems theory (e.g., attractor landscapes, change in attractors). Next, participants are divided into small groups (between 3 and 6 people). Each group is then provided with a description of a conflict case and instructed to deconstruct the case using the options offered by the software interface. First, the group members list factors that they agree are relevant to the case. Next, the group estimates each factor's (1) effect on momentary violence, (2) potential for positive long-term interactions (i.e., reinforcement of the positive attractor), and (3) potential for negative interactions (i.e., reinforcement of the negative attractor). These estimates are based on the discussions within the group. The agreed-upon effects are visualized on the monitor. The group then discusses how each factor appears to have affected the momentary state and attractors of the case. All participants in the workshop or class are then encouraged to ask practical and questions of the instructor.

After this introductory session, each group works on a complex conflict case, in which the short-term and the long-term consequences of various factors and actions may be very different. Some factors, for example, may have positive consequences in the short run (e.g., suppressing violence) but bad consequences in the long run (e.g., creating resentment). Conversely, some factors may have a negative effect in the short term but a positive effect in the long term. In some workshops, the participants discuss cases with which they are familiar or for which they have expertise. In an early workshop at Columbia University, for example, the case concerned an environmental conflict around Riverbank State Park in West Harlem, New York, which was constructed above a sewage treatment plant in New York City (Holloway, 1992). A short term solution that seemed to satisfy both the local city administration and the advocates for the local community was to build the sewage treatment plant, and to compensate the community by constructing a recreational area on the roof of the sewage treatment plant. This rooftop area was a large park (28 acres) with facilities allowing for sport and leisure activities. Although the solution seems integrative and satisfying in the short term, it becomes increasingly questionable when one considers long-term issues, such as potential health problems (children playing on the roof top of a building that processes 170 million gallons of raw sewage a day), noxious odor, and at the macro-level, discrimination, micro-violence, and racism.

In the next phase, participants move from the description of the case to generating possible transformation scenarios by employing the attractor software. Thus, subgroups of participants are encouraged to list factors and plan strategies aimed at (1) minimizing the negative attractor, (2) maximizing the positive attractor, and (3) moving the current state toward the positive attractor. In so doing, the participants assimilate key features of dynamical systems theory, and apply them directly to practical problems. The generated intervention scenarios prove remarkably mature, as participants consider both short-term and long-term consequences of the proposed changes with respect to conflict resolution and the creation of conditions for sustainable peace.

The final phase is dedicated to group discussion. Each subgroup discusses its case in the presence of all workshop or class members. The members indicate the factors they considered and the effectiveness of these factors as visualized by the attractor software. They also discuss their experience in using the software and their sense of the software's added value in understanding and resolving conflict. This phase of the workshop may prove critical in groups where different perspectives on the same problem are confronted because it affords them the opportunity to compare their results with the outcomes of different attractor software simulations. This allows for a debate concerning concrete issues, including long- versus short-term interests and potential changes resulting from proposed solutions. Beyond enabling the sharing of different perspectives, the visualization provides a space for each solution to be systematically considered.

Research Comparing Two Models of Negotiation Training

We investigated the effect of attractor software training on the sustainability of agreements negotiated by participants in two studies (Nowak et al., 2010a). Two groups were trained in integrative negotiation (Fisher, Ury, & Patton, 1991) and participated in a negotiation simulation game between a striking union and the management in a governmental organization. In addition to receiving integrative negotiations training, one group went through the attractor software tutorial before the negotiation game, while another (control) group received only the integrative negotiations training. Results show that compared to the members of the control group, members of the group that was trained with the attractor software demonstrated significantly greater ease in communication with their negotiation partner and better understanding of the negotiation process. Both effects become stronger over time (see the figure below).

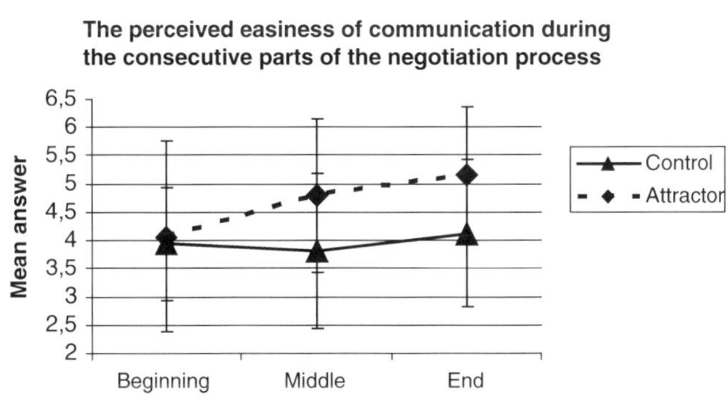

Relative Ease of Communication During Three Phases of the Negotiation Process

The most interesting findings, however, concern the outcomes of the negotiations (see figure, below). Although the groups did not reliably differ in their satisfaction with the negotiation process and the outcomes, nor in their perceptions of the fairness of the outcomes, there were reliable differences in the long-term stability of the negotiated final agreement. In fact, each group that negotiated with help of the attractor software achieved durable long-term solutions, whereas the majority of the other groups failed to do so.

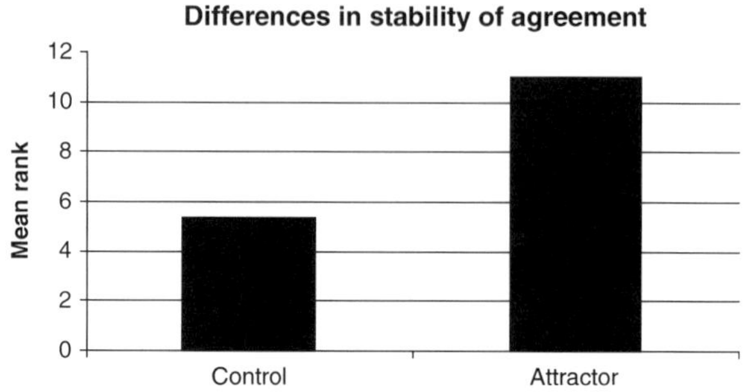

The Differences Between Groups According to the Stability of Reached Agreements

Although these studies are preliminary and did not control for the effects of additional training experienced by the attractor software groups, the results point to the tangible benefits of training with the attractor software on the sustainability of negotiated agreements. These findings also speak to the fact that even a highly satisfying outcome may not be durable and that this fact may go unnoticed to the parties until they are confronted with the consequences of their actions.

Potential Applications

Based on our experience employing the attractor software approach, we can point to three ways in which the software can be effectively utilized.

First, the interactive nature of the program allows the user to visualize how various factors impact both the momentary and long-term state of the relationship among the conflicting parties. This alerts the user to the distinction between interventions with immediate but no long-lasting effects and interventions with no immediately apparent effects but which target the attractor landscape, thereby creating new possibilities for lasting change.

Second, social science theory and data can be used to identify relevant factors in particular situations, and to define the impact of these factors on the whole system. The program can thus be used both to test the assumptions of existing social science data and to identify which factors should be considered. This could help researchers and practitioners identify theory-based attractors—both manifest and latent—in interpersonal and inter-group relations.

Third, users can interact with the software in small groups. Rather than arguing about general issues and positions, users can map the ensemble of relevant factors and test their respective assumptions and intuitions. In this way, common understanding can be achieved with respect to specific strategies for conflict resolution.

Finally, use of the software overcomes communication barriers, allowing for the exchange and synchronization of different mental models of a conflict. In this sense, it can contribute to the construction of a shared reality regarding a conflict that might otherwise be intractable.

References

Alba, R. D., & Nee, V. (2005). *Remaking the American mainstream: Assimilation and contemporary immigration*. Cambridge, MA: Harvard University Press.

Allport, G. W. (1954). *The nature of prejudice*. Reading, MA: Addison-Wesley.

Allport, G. W. (1969). *Becoming: Basic considerations for a psychology of personality*. New Haven, CT: Yale University Press.

Alvesson, M., & Willmott, H. (1992a). Critical theory and management studies: An introduction. In M. Alvesson & H. Willmott (Eds.), *Critical management studies* (pp. 1–20). London: Sage.

Alvesson, M., & Willmott, H. (1992b). On the idea of emancipation in management and organization studies. *The Academy of Management Review, 17*, 432–464.

Anderson, N. H. (1981). *Foundations of information integration theory*. New York, NY: Academic.

Anderson, R. M., & May, R. M. (1991). *Infectious diseases of humans: Dynamics and control*. New York, NY: Oxford University Press.

Archer, J. (2006). Testosterone and human aggression: An evaluation of the challenge hypothesis. *Neuroscience and Biobehavioral Reviews, 30*, 319–345.

Aron, A., & Corne, S. (1994). *Writings for a liberation of psychology*. Boston, MA: Harvard University Press.

Asch, S. E. (1946). Forming impressions of personalities. *Journal of Abnormal and Social Psychology, 41*, 258–290.

Asch, S. E. (1956). Studies of independence and conformity: A minority of one against a unanimous majority. *Psychological Monographs, 70*(9), 1–70. Whole No. 416.

Axelrod, R. (1984). *The evolution of cooperation*. New York, NY: Basic Books.

Axelrod, R., & Cohen, M. (2000). *Harnessing complexity: Organizational implications of a scientific frontier*. New York, NY: Basic Books.

Axelrod, R., Riolo, R. L., & Cohen, M. D. (2002). Beyond geography: Cooperation with persistent links in the absence of clustered neighborhoods. *Personality and Social Psychology Review, 6*, 341–346.

Azar, E. E. (1983). The theory of protracted social conflict and the challenge of transforming conflict situations. *Monograph Series in World Affairs, 20*(M2), 81–99.

Azar, E. E. (1986). Management of protracted social conflict in the third world. *Ethnic Studies Report, 4*(2), 1–17.

Azar, E. E. (1990). *The management of protracted social conflict*. Aldershot: Dartmouth.

Bak, P. (1997). *How nature works: The science of self-organized criticality*. New York, NY: Oxford University Press.

Baker-Brown, G., Ballard, E. J., Bluck, S., deVries, B., Suedfeld, P., & Tetlock, P. (1992). The integrative complexity coding manual. In C. Smith (Ed.), *Handbook of thematic analysis* (pp. 605–611). Cambridge, England: Cambridge University Press.

R.R. Vallacher et al., *Attracted to Conflict: Dynamic Foundations of Destructive Social Relations*, Peace Psychology Book Series, DOI 10.1007/978-3-642-35280-5,

Bales, R. F. (1950). *Interaction process analysis*. Chicago, IL: University of Chicago Press.

Barabasi, A. (2001). *Linked*. New York, NY: Penguin.

Bargh, J. A. (1996). Automaticity in social psychology. In E. T. Higgins & A. W. Kruglanski (Eds.), *Social psychology: Handbook of basic principles* (pp. 169–183). New York, NY: Guilford Press.

Baron, R. M., Amazeen, P. G., & Beek, P. J. (1994). Local and global dynamics of social relations. In R. R. Vallacher & A. Nowak (Eds.), *Dynamical systems in social psychology* (pp. 111–138). San Diego, CA: Academic.

Barry, B., & Oliver, R. L. (1996). Affect in dyadic negotiation: A model and propositions. *Organizational Behavior and Human Decision Processes, 70*, 175–187.

Bar-Tal, D. (1998). Societal beliefs in times of intractable conflict: The Israeli case. *International Journal of Conflict Management, 9*, 22–50.

Bar-Tal, D. (2000). From intractable conflict through conflict resolution to reconciliation: Psychological analysis. *Political Psychology, 21*(2), 351–365.

Bar-Tal, D. (2007). Social-psychological foundations of intractable conflict. *The American Behavioral Scientist, 50*(11), 1430–1453.

Bar-Tal, D., Raviv, A., & Freund, T. (1994). An anatomy of political beliefs: A study of their centrality, contents, and epistemic authority. *Journal of Applied Social Psychology, 24*, 849–872.

Bartoli, A., Bui-Wrzosinska, L., & Nowak, A. (2010). Peace is in movement: A dynamical systems perspective on the emergence of peace in Mozambique. *Peace and Conflict: Journal of Peace Psychology, 16*(2), 211–230.

Baumeister, R. F., Bratslavsky, E., Muraven, M., & Tice, D. M. (1998). Ego depletion: Is the active self a limited resource? *Journal of Personality and Social Psychology, 74*, 1252–1265.

Baumeister, R. F., & Heatherton, T. D. (1996). Self-regulation failure: An overview. *Psychological Inquiry, 7*, 1–15.

Baumeister, R. F., Heatherton, T. F., & Tice, D. M. (1994). *Losing control: How and why people fail at self-regulation*. San Diego, CA: Academic.

Baumeister, R. F., Vohs, K. D., & Tice, D. M. (2007). The strength model of self-control. *Current Directions in Psychological Science, 16*, 351–355.

Bechtoldt, M. N., De Dreu, C. K. W., Nijstad, B. A., & Choi, H. S. (2010). Motivated information processing, social tuning, and group creativity. *Journal of Personality and Social Psychology, 99*, 622–637.

Beek, P. J., & Hopkins, B. (1992). Four requirements for a dynamical systems approach to the development of social coordination. *Human Movement Science, 11*, 425–442.

Bennett, D. S. (1996). Security, bargaining, and the end of interstate rivalry. *International Studies Quarterly, 40*, 157–184.

Bercovitch, J. (2005). Mediation in the most resistant cases. In C. A. Crocker, F. O. Hampson, & P. Aall (Eds.), *Grasping the nettle: Analyzing cases of intractable conflict*. Washington, DC: United States Institute of Peace.

Berry, J. S. (1997). Immigration, acculturation, and adaptation. *Applied Psychology, 46*(1), 5–34.

Blake, R. R., Mouton, J. S., Barnes, L. B., & Greiner, L. E. (1964). Breakthrough in organization development. *Harvard Business Review, 42*(6), 133–155.

Blalock, H. M. (1989). *Power and conflict: Toward a general theory*. Newbury Park, CA: Sage.

Blum, G. (2007). *Islands of agreement: Managing enduring armed rivalries*. Cambridge, MA: Harvard University Press.

Boardman, S., & Horowitz, S. (1994). Constructive conflict management and social problems: An introduction. *Journal of Social Issues, 50*(1), 1–12.

Brehm, S. S., & Brehm, J. W. (1981). *Psychological reactance: A theory of freedom and control*. New York, NY: Academic.

Brewer, M. B. (1979). In-group bias in the minimal intergroup situation: A cognitive-motivational analysis. *Psychological Bulletin, 86*, 307–324.

Brewer, M. B., & Brown, R. J. (1998). Intergroup relations. In D. T. Gilbert, S. T. Fiske, & G. Lindzey (Eds.), *The handbook of social psychology* (4th ed., Vol. 2, pp. 554–594). New York, NY: McGraw-Hill.

Brown, C. T., & Liebovitch, L. S. (2010). *Fractal analysis: Quantitative applications in the social sciences* (Vol. 165). Thousand Oaks, CA: Sage.

Brunswick, E. (1956). *Perception and the representative design of experiments* (2nd ed.). Berkeley: University of California Press.

Burgess, H., & Burgess, G. (1996). Constructive confrontation: A transformative approach to intractable conflicts. *Mediation Quarterly, 13*(4), 305–322.

Burton, J. (1987). *Resolving deep-rooted conflict: A handbook*. Lanham, MD: University Press of America.

Cacioppo, J. T., & Bernston, G. G. (1994). Relationship between attitudes and evaluative space: A critical review with emphasis on the separability of positive and negative substrates. *Psychological Bulletin, 15*, 401–423.

Cairns, E., & Darby, J. (1998). The conflict in Northern Ireland: Causes, consequences, and controls. *The American Psychologist, 53*, 754–760.

Campbell, D. T. (1965). *Ethnocentric and other altruistic motives*. Lincoln, NE: University of Nebraska Press.

Canary, D. J., Cupach, W. R., & Messman, S. J. (1995). *Relationship conflict: Conflict in parent–child, friendship, and romantic relationships (series on close relationships, Clyde and Susan Hendrick, series editors)*. Newbury Park, CA: Sage.

Carnevale, P. J., & Probst, T. (1998). Social values and social conflict in creative problem solving and categorization. *Journal of Personality and Social Psychology, 74*, 1300–1309.

Cartwright, D. (1979). Contemporary social psychology in historical perspective. *Social Psychology Quarterly, 42*, 82–93.

Carver, C. S., & Scheier, M. F. (1999a). Stress, coping and self-regulatory processes. In L. Pervin & J. P. Oliver (Eds.), *Handbook of personality: Theory and research* (pp. 553–575). New York, NY: Guilford Press.

Carver, C. S., & Scheier, M. F. (1999b). Themes and issues in the self-regulation of behavior. In R. S. Wyer Jr. (Ed.), *Advances in social cognition* (Vol. 12, pp. 1–105). Mahwah, NJ: Erlbaum.

Christakis, N. A., & Fowler, J. H. (2009). *Connected: The surprising power of our social networks and how they shape our lives*. New York, NY: Little, Brown.

Coleman, P. T. (2000a). Intractable conflict. In M. Deutsch & P. T. Coleman (Eds.), *The handbook of conflict resolution: Theory and practice* (pp. 428–450). San Francisco, CA: Jossey Bass.

Coleman, P. T. (2000b). Fostering ripeness in seemingly intractable conflict: An experimental study. *International Journal of Conflict Management, 11*(4), 300–317.

Coleman, P. T. (2003). Characteristics of protracted, intractable conflict: Toward the development of a metaframework—I. *Peace and Conflict: Journal of Peace Psychology, 9*(1), 1–37.

Coleman, P. T. (2004). Paradigmatic framing of protracted, intractable conflict: Towards the development of a meta-framework—II. *Peace and Conflict: Journal of Peace Psychology, 10*(3), 197–235.

Coleman, P. T. (2006). Conflict, complexity, and change: A meta-framework for addressing protracted, intractable conflicts—III. *Peace and Conflict: Journal of Peace Psychology, 12*, 325–348.

Coleman, P. T. (2011). *The five percent: Finding solutions to seemingly impossible conflicts*. New York, NY: Public Affairs.

Coleman, P. T. (2012). The essence of peace? toward a comprehensive and parsimonious model of sustainable peace. In P. T. Coleman & M. Deutsch (Eds.), *Psychological contributions to sustainable peace*. New York, NY: Springer.

Coleman, P. T., Bui-Wrzosinska, L., Vallacher, R. R., & Nowak, A. (2006). Protracted conflicts as dynamical systems. In A. K. Schneider & C. Honeyman (Eds.), *The negotiator's fieldbook: The desk reference for the experienced negotiator* (pp. 61–74). Chicago: American Bar Association.

Coleman, P. T., & Deutsch, M. (2001). Introducing cooperation and conflict resolution into schools: A systems approach. In D. J. Christie, R. V. Wagner, & D. D. N. Winter (Eds.), *Peace, conflict and violence: Peace psychology for the 21st century*. Upper Saddle River, NJ: Prentice-Hall.

Coleman, P. T., & Deutsch, M. (Eds.). (2012). *Psychological components of sustainable peace*. New York, NY: Springer.

Coleman, P. T., Hacking, A., Stover, M., Fisher-Yoshida, B., & Nowak, A. (2008). Reconstructing ripeness I: A study of constructive engagement in protracted social conflicts. *Conflict Resolution Quarterly, 26*(1), 3–42.

Coleman, P. T., & Lim, J. (2001). A systematic approach to evaluating the effects of collaborative negotiation training on individuals and groups. *Negotiation Journal, 17*(4), 363–392.

Coleman, P. T., Vallacher, R. R., Bartoli, A., Nowak, A., & Bui-Wrzosinska, L. (2011). Navigating the landscape of conflict: Applications of dynamical systems theory to addressing protracted conflict. In D. Korppen, N. Ropers, & H. J. Giessmann (Eds.), *The non-linearity of peace processes: Theory and practice of systemic conflict transformation* (pp. 39–56). Opladen, Germany: Barbara Budrich.

Coleman, P. T., Vallacher, R., Nowak, A., & Bui-Wrzosinska, L. (2007). Intractable conflict as an attractor: Presenting a dynamical model of conflict, escalation, and intractability. *The American Behavioral Scientist, 50*, 1454–1475.

Collier, P. (2000). *Economic causes of civil conflict and their implications for policy*. Washington, DC: World Bank.

Comas-Díaz, L., Lykes, M. B., & Alarcón, R. (1998). Ethnic conflict and psychology of liberation in Guatemala, Perú, and Puerto Rico. *The American Psychologist, 53*, 778–792.

Condon, W. S., & Ogston, W. D. (1967). A segmentation of behavior. *Journal of Psychiatric Research, 5*, 221–235.

Cooley, C. H. (1902). *Human nature and the social order*. New York: Scribner.

Costa, P. T., & McCrae, R. R. (1995). Domains and facets: Hierarchical personality assessment using the revised NEO personality inventory. *Journal of Personality Assessment, 64*, 21–50.

Crocker, C. A., & Hampson, F. O. (2004). *Taming intractable conflicts: Mediation in the hardest cases*. Washington, DC: United States Institute of Peace.

Crocker, C. A., Hampson, F. O., & Aall, P. (1999). *Herding cats: Multiparty mediation in a complex world*. Washington, DC: United States Institute of Peace.

Crocker, C. A., Hampson, F. O., & Aall, P. (2005). *Grasping the nettle: Analyzing cases of intractable conflict*. Washington, DC: United States Institute of Peace.

Crocker, J., & Luhtanen, R. (1990). Collective self-esteem and ingroup bias. *Journal of Personality and Social Psychology, 58*, 60.

Dabs, J. M., Jr., & Morris, R. (1990). Testosterone, social class, and antisocial behavior in a sample of 4462 men. *Psychological Science, 1*, 209–211.

Daly, M., & Wilson, M. (1988). *Homicide*. New York, NY: Aldine deGruyter.

Daly, M., & Wilson, M. (1994). Evolutionary psychology of male violence. In J. Archer (Ed.), *Male violence* (pp. 253–288). New York, NY: Routledge.

Darwin, C. (1859). *On the origin of species by means of natural selection, or the preservation of favoured races in the struggle for life*. London: John Murray.

Darwin, C. (1871). *On the origin of species* (5th ed.). London: John Murray.

Davis, B. P., & Knowles, E. S. (1999). A disrupt-then-reframe technique of social influence. *Journal of Personality and Social Psychology, 76*, 192–199.

Davis-Stober, C. P., Dana, J., & Budescu, D. V. (2010). Why recognition is rational: Optimality results on single-variable decision rules. *Judgment and Decision Making, 5*(4), 216–229.

De Dreu, C. K. W. (2010). Social conflict: The emergence and consequences of struggle and negotiation. In S. T. Fiske, D. T. Gilbert, & G. Lindzey (Eds.), *Handbook of social psychology* (5th ed., pp. 983–1023). San Francisco, CA: Wiley.

De Dreu, C. K. W., Beersma, B., Steinel, W., & Van Kleef, G. A. (2007). The psychology of negotiation: Principles and basic processes. In A. W. Kruglanski & E. T. Higgins (Eds.), *Social psychology: Handbook of basic principles* (2nd ed.). New York, NY: Guilford.

De Dreu, C. K. W., Beersma, B., Stroebe, K., & Euwema, M. C. (2006). Motivated information processing, strategic choice, and the quality of negotiated agreement. *Journal of Personality and Social Psychology, 90*, 927–943.

De Dreu, C. K. W., & Gelfand, M. J. (2008). Conflict in the workplace: Sources, dynamics and functions across multiple levels of analysis. In C. K. W. De Dreu & M. J. Gelfand (Eds.), *The psychology of conflict and conflict management in organizations*. New York, NY: Lawrence Erlbaum.

De Dreu, C. K., & Van Lange, P. A. (1995). The impact of social value orientations on negotiator cognition and behavior. *Personality and Social Psychology Bulletin, 21*, 1178–1188.

De Dreu, C. K. W., & Weingart, L. R. (2003). Task versus relationship conflict, team performance, and team member satisfaction: A meta-analysis. *The Journal of Applied Psychology, 88*, 741–749.

De Dreu, C. K. W., Weingart, L. R., & Kwon, S. (2000). Influence of social motives on integrative negotiation: A meta-analytical review and test of two theories. *Journal of Personality and Social Psychology, 78*, 889–905.

De Wit, F. R. C., Greer, L. L., & Jehn, K. A. (2011). The paradox of intragroup conflict: A meta-analysis. *The Journal of Applied Psychology, 97*(2), 360–390.

De Zavala, A. G., Cichocka, A., Eidelson, R., & Jayawickreme, N. (2009). Collective narcissism and its social consequences. *Journal of Personality and Social Psychology, 97*, 1074.

DeCharms, R., & Moeller, G. H. (1962). Values expressed in American children's readers. *Journal of Abnormal and Social Psychology, 64*, 136–142.

Dechesne, M., Greenberg, J., Arndt, J., & Schimel, J. (2000). Terror management and the vicissitudes of sports fan affiliation: The effects of mortality salience on optimism and fan identification. *European Journal of Social Psychology, 30*, 813–835.

Deutsch, M. (1949a). An experimental study of the effects of co-operation and competition upon group processes. *Human Relations, 2*, 199–231. Retrieved from http://psycnet.apa.org/psycinfo/1950-04051-001.

Deutsch, M. (1949b). The directions of behavior: A field-theoretic approach to the understanding of inconsistencies. *Journal of Social Issues, 25*(1), 43–51.

Deutsch, M. (1969). Conflicts: Productive and destructive. *Journal of Social Issues, 25*(1), 7–42.

Deutsch, M. (1973). *The resolution of conflict: Constructive and destructive processes*. New Haven, CT: Yale University Press.

Deutsch, M. (1977). *The resolution of conflict: Constructive and destructive processes*. New Haven: Yale University Press.

Deutsch, M. (1985). *Distributive justice: A social psychological perspective*. New Haven, CT: Yale University Press.

Deutsch, M. (2002). Social psychology's contributions to the study of conflict resolution. *Negotiation Journal, 18*(4), 307–320.

Deutsch, M. (2006a). A framework for thinking about oppression and its change. *Social Justice Research, 19*, 7–41.

Deutsch, M. (2006b). Cooperation and competition. In M. Deutsch, P. T. Coleman, & E. C. Marcus (Eds.), *The handbook of conflict resolution: Theory and practice* (pp. 23–42). San Francisco, CA: Wiley.

Deutsch, M., & Coleman, P. T. (2000). *The handbook of conflict resolution: Theory and practice*. San Francisco, CA: Jossey-Bass.

Deutsch, M., Coleman, P. T., & Marcus, E. C. (Eds.). (2006). *Handbook of conflict resolution: Theory and practice* (2nd ed.). San Francisco, CA: Jossey-Bass.

Deutsch, M., & Gerard, H. B. (1955). A study of normative and informational social influence upon individual judgment. *Journal of Abnormal and Social Psychology, 51*, 629–636.

Diamond, L., & McDonald, J. (1996). *Multi-track diplomacy: A systems approach to peace*. West Hartford, CT: Kumarian Press.

Diehl, P. F., & Goertz, G. (2001). *War and peace in international rivalry*. Ann Arbor, MI: University of Michigan Press.

Diesing, P. (1962). *Reason in society: Five types of decisions and their social conditions*. Urbana, IL: University of Illinois Press.

Dijksterhuis, A., & Nordgren, L. F. (2006). A theory of unconscious thought. *Perspectives on Psychological Science, 1*, 95–109.

Dittman, A. T., & Llewellyn, L. G. (1969). Body movement and speech rhythm in social conversation. *Journal of Personality and Social Psychology, 11*, 98–106.

Dörner, D. (1996). *The logic of failure: Why things go wrong and what we can do to make them right*. New York, NY: Holt.

Dovidio, J. F., & Gaertner, S. L. (1986). *Prejudice, discrimination, and racism*. New York, NY: Academic.

Druckman, D. (1994). Determinants of compromising behavior in negotiation: A meta-analysis. *Journal of Conflict Resolution, 38*, 507–556.

Druckman, D. (2001). Turning points in international negotiation: A comparative analysis. *Journal of Conflict Resolution, 45*, 519–544.

Druckman, D. (2007). Negotiating in the international context. In I. W. Zartman (Ed.), *Peacemaking in international conflict methods and techniques* (pp. 111–162). New York: United States Inst of Peace.

Durkheim, E. (1938). *The rules of sociological method*. Chicago: University of Chicago Press.

Ellis, L. (1986). Evidence of neuroandrogenic etiology of sex roles form a combined analysis of human, non-human primate and nonprimate mammalian studies. *Personality and Individual Differences, 7*, 519–552.

Emerson, R. M. (1962). Power-dependence relations. *American Sociological Review, 27*, 31–41.

Fiol, C. M., Pratt, M. G., & O'Connor, E. J. (2009). Managing intractable identity conflicts. *The Academy of Management Review, 34*(1), 32–55.

Fisher, R. (1994). Generic principals for resolving intergroup conflict. *Journal of Social Issues, 50* (1), 47–68.

Fisher, R. J., & Keashly, L. (1990). A contingency approach to third party intervention. In R. J. Fisher (Ed.), *The social psychology of intergroup and international conflict resolution* (pp. 234–238). New York: Springer.

Fisher, R. J., & Keashly, L. (1991). The potential complementarity of mediation and consultation within a contingency model of Third Party Intervention. *Journal of Peace Research, 28*, 29–42.

Fisher, R., Ury, W., & Patton, B. (1991). *Getting to yes: Negotiating an agreement without giving in*. New York, NY: Houghton Mifflin Books.

Fiske, S. (1998). Stereotyping, prejudice, and discrimination. In D. Gilbert, S. Fiske, & G. Lindzey (Eds.), *The handbook of social psychology* (Vol. 2, pp. 357–411). New York: McGraw-Hill.

Fiske, S. (2000). Stereotyping, prejudice, and discrimination at the seam between the centuries: Evolution, culture, mind, and brain. *European Journal of Social Psychology, 29*, 188–211.

Fiske, S. (2002). What we know now about bias and intergroup conflict, the problem of the century. *Current Directions in Psychological Science, 11*, 123–128.

Fiske, S. T., & Taylor, S. E. (1991). *Social cognition* (2nd ed.). New York, NY: Random House.

Follet, M. P. (1973). Power. In E. M. Fox & L. Urwick (Eds.), *Dynamic administration: The collected papers of Mary Parker Follett* (pp. 66–87). London: Pitman. Original work published in 1925.

Fredrickson, B. L. (1998). What good are positive emotions? *Review of General Psychology, 2*, 300–319.

Fredrickson, B. L. (2001). The role of positive emotions in positive psychology: The broaden-and-build theory of positive emotions. *The American Psychologist, 56*, 218–226.

Fredrickson, B. L. (2006). Unpacking positive emotions: Investigating the seeds of human flourishing. *The Journal of Positive Psychology, 1*, 57–59.

Freud, S. (1920/1964). *Beyond the pleasure principle (J. Strachey, Trans.)*. New York, NY: Norton.

Freud, S. (1936). *The problem of anxiety*. New York, NY: W. W. Norton.

Freud, S. (1937). *The ego and the mechanisms of defense*. London: Hogarth.

Freud, S. (1960). The psychopathology of everyday life. In J. Strachey (Ed.), *The standard edition of the complete psychological works of Sigmund Freud (Tyson, A. Trans.)* (Vol. 6). London: Hogarth Press. Original work published 1901.

Fry, D. P. (2006). *The human potential for peace: An anthropological challenge to assumptions about war and violence*. New York, NY: Oxford University Press.

Fry, D. P., & Miklikowska, M. (2012). Culture of peace. In P. T. Coleman & M. Deutsch (Eds.), *Psychological contributions to sustainable peace*. New York, NY: Springer.

Gailliot, M. T., Baumeister, R. F., DeWall, C. N., Maner, J. K., Plant, E. A., Tice, D. M., et al. (2007). Self-control relies on glucose as a limited energy source: Willpower is more than a metaphor. *Journal of Personality and Social Psychology, 92*, 325–336.

Gerhart, B., & Rynes, S. L. (1991). Determinants and consequences of salary negotiations by male and female MBA graduates. *The Journal of Applied Psychology, 76*, 256–262.

Gersick, C. J. G. (1991). Revolutionary change theories: A multilevel exploration of the punctuated equilibrium paradigm. *The Academy of Management Review, 16*, 10–36.

Gilbert, D. T. (1989). Thinking lightly about others: Automatic components of the social inference process. In J. S. Uleman & J. A. Bargh (Eds.), *Unintended thought* (pp. 189–211). New York, NY: Guilford Press.

Gilbert, D. T. (1993). The assent of man: Mental representation and the control of belief. In D. M. Wegner & J. W. Pennebaker (Eds.), *The handbook of mental control* (pp. 57–87). Englewood Cliffs, NJ: Prentice Hall.

Gilbert, D. T., & Hixon, J. G. (1991). The trouble of thinking: Activation and applications of stereotypical beliefs. *Journal of Personality and Social Psychology, 60*, 509–517.

Gilbert, D. T., Pelham, B. W., & Krull, D. S. (1988). On cognitive busyness: When person perceivers meet persons perceived. *Journal of Personality and Social Psychology, 54*, 733–740.

Glass, D. C., & Singer, J. E. (1972). *Urban stress: Experiments on noise and social stressors*. New York, NY: Academic.

Gleick, P. H. (1993). Water and conflict: Fresh water resources and international security. *International Security, 18*(1), 79–112.

Goertz, G., & Diehl, P. F. (1992). The empirical importance of enduring rivalries. *International Interactions, 18*(2), 151–163.

Goertz, G., & Diehl, P. F. (1993). Enduring rivalries: Theoretical constructs and empirical patterns. *International Studies Quarterly, 37*, 147–171.

Gottman, J. M., Markman, H., & Notarius, C. (1977). The topography of marital conflict: A sequential analysis of verbal and nonverbal behavior. *Journal of Marriage and the Family, 39*, 461–477.

Gottman, J. M., Murray, J. D., Swanson, C. C., Tyson, R., & Swanson, K. R. (2002). *The mathematics of marriage*. Cambridge, MA: MIT Press.

Gottman, J. M., & Silver, N. (1999). *The seven principles for making marriage work*. New York, NY: Three Rivers Press.

Gottman, J., Swanson, C., & Swanson, K. (2002). A general systems theory of marriage: Nonlinear difference equation modeling of marital interaction. *Personality and Social Psychology Review, 6*, 326–340.

Gray, B., Coleman, P. T., & Putnam, L. L. (2007). Introduction: Intractable conflict: new perspectives on the causes and conditions for change. *The American Behavioral Scientist, 50*, 1415–1429.

Greenberg, J., Porteus, J., Simon, L., Pyszczynski, T., & Solomon, S. (1995). Evidence of a terror management function of cultural icons: the effects of mortality salience on the inappropriate use of cherished cultural symbols. *Personality and Social Psychology Bulletin, 21*, 1221–1228.

Greenberg, J., Pyszczynski, T., Solomon, S., Rosenblatt, A., Veeder, M., Kirkland, S., et al. (1990). Evidence for terror management theory II: The effects of mortality salience on reactions to those who threaten or bolster the cultural worldview. *Journal of Personality and Social Psychology, 58*, 308–318.

Greenwald, A. G., & Banaji, M. R. (1995). Implicit social cognition: Attitudes, self-esteem, and stereotypes. *Psychological Review, 102*, 4–27.

Guastello, S. (1995). *Chaos, catastrophe, and human affairs: Application of nonlinear dynamics to work, organizations, and social evolution*. Mahwah, NJ: Lawrence Erlbaum Associates.

Guastello, S., Koopmans, M., & Pincus, D. (Eds.). (2009). *Chaos and complexity in psychology: The theory of nonlinear dynamical systems*. New York, NY: Cambridge University Press.

Guastello, S. J., Pincus, D., & Gunderson, P. R. (2006). Electrodermal arousal between participants in a conversation: Nonlinear dynamics and linkage effects. *Nonlinear Dynamics, Psychology, and Life Sciences, 10*, 365–399.

Guckenheimer, J., & Holmes, P. (1983). *Nonlinear oscillations, dynamical systems, and bifurcations of vector fields*. New York, NY: Springer.

Gurr, T. R. (2000). *Peoples versus states*. Washington, DC: United States Institute of Peace Press.

Gurr, T. R., & Scarritt, J. R. (1989). Minorities rights at risk: A global survey. *Human Rights Quarterly, 11*, 375.

Haken, H. (1978). *Synergetics*. Berlin, Germany: Springer.

Higgins, E. T. (1997). Beyond pleasure and pain. *The American Psychologist, 52*(12), 1280–1300.

Hinkle, S., & Brown, R. (1990). Intergroup comparisons and social identity: Some links and lacunae. *Social identity theory: Constructive and critical advances, 48*, 70–86.

Hogg, M. A., & Williams, K. D. (2000). From I to we: Social identity and the collective self. *Group dynamics: Theory, research, and practice, 4*(1), 81–97.

Holland, J. (1975). *Adaptation in natural and artificial systems*. Ann Arbor: The University of Michigan Press.

Holland, J. (1995a). Can there be a unified theory of complex adaptive systems? In H. J. Morowitz & J. L. Singer (Eds.), *The mind, the brain, and complex adaptive systems*. Boston, MA: Addison-Wesley.

Holland, J. H. (1995b). *Emergence: From chaos to order*. Reading, MA: Addison-Wesley.

Holloway, L. (1992). 28 Acres of roof and a place to play in west harlem; riverbank state park is under construction atop north river sewage treatment plant. *New York Times*, p. 1.

James, W. (1890). *Principles of psychology*. New York, NY: Holt.

Jehn, K. A. (1995). A multimethod examination of the benefits and detriments of intragroup conflict. *Administrative Science Quarterly, 40*, 256–282.

Jehn, K. A. (1997). A qualitative analysis of conflict types and dimensions in organizational groups. *Administrative Science Quarterly, 42*(3), 530–557.

Johnson, S. (2001). *Emergence: The connected lives of ants, brains, cities, and software*. New York, NY: Scribner.

Johnson, D. W., & Johnson, R. T. (1989). *Cooperation and competition: Theory and research*. Edina, MN: Interaction Book.

Johnson, D. W., & Johnson, R. T. (1994). *Learning together and alone. Cooperative, competitive, and individualistic learning*. Boston, MA: Allyn & Bacon.

Johnson, D. W., & Johnson, R. T. (2005). New developments in social interdependence theory. *Psychology Monograph, 13*, 285–360.

Johnson, D. W., & Johnson, R. T. (2006). Peace education for consensual peace: The essential role of conflict resolution. *Journal of Peace Education, 3*(2), 147–174.

Johnson, D. W., Johnson, R. T., & Smith, K. A. (2000). Constructive controversy: The educative power of intellectual conflict. *Change: The Magazine of Higher Learning, 32*(1), 28–37.

Jones, E. E. (1985). Major developments in social psychology during the past five decades. *Handbook of Social Psychology, 1*, 47–107.

Jones, E. E. (1998). Major developments in five decades of social psychology. In D. T. Gilbert, S. T. Fiske, & G. Lindzey (Eds.), *The handbook of social psychology* (4th ed.). New York, NY: McGraw-Hill.

Jones, M. (2006). *The American pursuit of unhappiness: Gross National Happiness (GNH), a new socioeconomic policy. Executive white paper.* Las Vegas, NV: International Institute of Management.

Jones, E. E., & Gerard, H. B. (1967). *Foundations of social psychology.* New York, NY: Wiley.

Jones, W., & Hughes, S. H. (2003). Complexity, conflict resolution and how the mind works. *Conflict Resolution Quarterly, 20,* 4–20.

Jost, J. T., & Banaji, M. R. (1994). The role of stereotyping in system-justification and the production of false consciousness. *British Journal of Social Psychology, 33,* 1–27.

Karpf, F. B. (1932). *American social psychology: Its origins, development, and European background.* New York, NY: Mcgraw-Hill Book. Retrieved from http://psycnet.apa.org/psycinfo/2006-04433-000/.

Katz, D., & Kahn, R. L. (1978). *The social psychology of organizations* (2nd ed.). New York, NY: Wiley.

Kauffman, S. (1995). *At home in the universe.* New York, NY: Oxford University Press.

Kegan, R. (1994). *In over our heads: The mental demands of modern life.* Cambridge, MA: Harvard University Press.

Kelley, H. H. (1967). Attribution theory in social psychology. In D. Levine (Ed.), *Nebraska symposium on motivation.* Lincoln, NE: University of Nebraska Press.

Kelley, H. H., & Thibaut, J. W. (1978). *Interpersonal relations: A theory of interdependence.* New York, NY: Wiley.

Kelman, H. C. (1997). Social-psychological dimensions of international conflict. In I. W. Zartman & J. L. Rasmussen (Eds.), *Peacemaking in international conflict: Methods and techniques* (pp. 191–237). Washington, DC: United States Institute of Peace Press.

Kelman, H. (1999). The interdependence of Israeli and Palestinian national identities: The role of the other in existential conflicts. *Journal of Social Issues, 55*(3), 581–600.

Kelso, J. A. S. (1995). *Dynamic patterns: The self-organization of brain and behavior.* Cambridge, MA: MIT Press.

Kim, P. H. (1997). Strategic timing in group negotiations: The implications of forced entry and forced exit for negotiators with unequal power. *Organizational Behavior and Human Decision Processes, 71,* 263–286.

Kim, P. H., & Fragale, A. R. (2005). Choosing the path to bargaining power: An empirical comparison of BATNAs and contributions in negotiations. *The Journal of Applied Psychology, 90,* 373–381.

Klein, K. J., Goertz, G., & Diehl, P. (2008). The peace scale: Conceptualizing and operationalizing nonrivalry and peace. *Conflict Management and Peace Science, 25,* 67–80.

Klein, K. J., & Kozlowski, S. W. J. (Eds.). (2000). *Multilevel theory, research, and methods in organizations: Foundations, extensions, and new directions* (Society for Industrial and Organizational Psychology Frontiers Series, pp. 467–511). San Francisco, CA: Jossey-Bass.

Koo, J., Han, J., & Kim, J. (2002). Integrative complexity of South–North Korean correspondences: A time-series analysis, 1984–1994. *Journal of Conflict Resolution, 46,* 286–304. doi:10.1177/0022002702046002006.

Koo, J., & Kim, J. (1999). Integrative complexity in newspaper columns and the bulletins of political parties: The impact of economic crisis and political transition on integrative complexity. *Korean Journal of Social & Personality Psychology, 13,* 35–52.

Kozlowski, S. W. J., & Klein, K. J. (2000). *A multilevel approach to theory and research in organizations: Contextual, temporal, and emergent processes.* San Francisco, CA: Jossey-Bass.

Krech, D., & Crutchfield, R. S. (1948). *Theory and problems of social psychology.* New York, NY: McGraw-Hill.

Kressel, K. (2006). Mediation revisited. In M. Deutsch, P. T. Coleman, & E. Marcus (Eds.), *The handbook of conflict resolution: Theory and practice*. San Francisco, CA: Jossey Bass.
Kriesberg, L. (1999). Intractable conflict. In E. Weiner (Ed.), *The handbook of interethnic coexistence* (pp. 332–342). New York, NY: Continuum.
Kriesberg, L. (2003). *Constructive conflicts: From escalation to resolution* (2nd ed.). Lanham, MD: Rowman and Littlefield.
Kriesberg, L. (2005). Nature, dynamics, and phases of intractability. In C. A. Crocker, F. O. Hampson, & P. Aall (Eds.), *Grasping the nettle: Analyzing cases of intractable conflict* (pp. 65–98). Washington, DC: United States Institute of Peace.
Kriesberg, L., Northrup, T. A., & Thorson, S. J. (1989). *Intractable conflicts and their transformation*. Syracuse, NY: Syracuse University Press.
Kruglanski, A. W., & Webster, D. M. (1996). Motivated closing of the mind: "Seizing" and "freezing.". *Psychological Review, 103*, 263–283.
Kugler, K. G., Coleman, P. T., & Fuchs, A. M. (2011). *Conflict, complexity, and openness: constructive* versus *destructive dynamics of discussions over intractable issues* (WOP Working Paper No. 2011/3). Retrieved from http://www.psy.lmu.de/wirtschaftspsychologie/forschung/working_papers/index.html
Kuhn, T. S. (1970). *The structure of scientific revolutions* (2nd ed.). Chicago, IL: University of Chicago Press.
Latané, B. (1981). The psychology of social impact. *The American Psychologist, 36*, 343–356.
Latané, B., & Nowak, A. (1994). Attitudes as catastrophes: From dimensions to categories with increasing involvement. In R. R. Vallacher & A. Nowak (Eds.), *Dynamical systems in social psychology* (pp. 219–249). San Diego, CA: Academic.
Latané, B., & Nowak, A. (1997). The causes of polarization and clustering in social groups. *Progress in Communication Sciences, 13*, 43–75.
Latané, B., Nowak, A., & Liu, J. H. (1994). Measuring emergent social phenomena: Dynamism, polarization, and clustering as order parameters of social systems. *Behavioral Science, 39*(1), 1–24.
Latane', B., & Nowak, A. (1994). Attitudes as catastrophes: From dimensions to categories with increasing involvement. In R. R. Vallacher & A. Nowak (Eds.), *Dynamical systems in social psychology* (pp. 219–249). San Diego, CA: Academic.
Lederach, J. P. (1997). *Building peace: Sustainable reconciliation in divided societies*. Washington, DC: United States Institute of Peace.
Levy, J. S. (1996). Contending theories of international conflict: A levels-of-analysis approach. In C. A. Crocker, F. O. Hampson, & P. Aall (Eds.), *Managing global chaos: Sources of and responses to international conflict*. Washington, DC: United States Institute of Peace.
Lewenstein, M., Nowak, A., & Latané, B. (1993). Statistical mechanics of social impact. *Physics Review A, 45*, 703–716.
Lewicki, R., Literer, A. J., Minton, J., & Saundars, D. M. (2004). *Negotiation* (3rd ed.). Boston, MA: Irwin/McGraw-Hill.
Lewicki, R. J., Saunders, D. M., & Minton, J. W. (1999). *Negotiation* (3rd ed.). New York, NY: McGraw-Hill.
Lewin, K. (1935). *A dynamic theory of personality*. New York, NY: McGraw-Hill.
Lewin, K. (1936). *Principles of topological psychology*. New York, NY: McGraw-Hill.
Lewin, K. (1938). *The conceptual representation and measurement of psychological forces*. Durham, NC: Duke University Press.
Lewin, K. (1943). Defining the "Field at a Given Time". *Psychological Review, 50*, 292–310. Republished in resolving social conflicts & field theory in social science, Washington, DC.
Lewin, K. (1947). Frontiers in group dynamics. *Human Relations, 1*, 5–41.
Lewin, K. (1948a). *Field theory in social sciences*. New York, NY: Harper & Brothers.
Lewin, K. (1948b). *Resolving social conflicts*. New York, NY: Harper & Brothers.

Liebovitch, L. S., Naudot, V., Vallacher, R., Nowak, A., Bui-Wrzosinska, L., & Coleman, P. T. (2008). Dynamics of two-actor cooperation–competition conflict models. *Physica A, 387*, 6360–6378.

Liebovitch, L. S., & Scheurle, D. (2000). Two lessons from fractals and chaos. *Complexity, 5*(4), 34–43.

Liebovitch, L. S., Vallacher, R., & Michaels, J. (2010). Dynamics of cooperation-competition interaction model. *Peace and Conflict: The Journal of Peace Psychology, 16*(2), 175–188.

Liht, J., Suedfeld, P., & Krawczyk, A. (2005). Integrative complexity in face-to-face negotiations between the Chiapas Guerrillas and the Mexican government. *Political Psychology, 26*, 543–552.

Livingstone, A., & Haslam, S. A. (2008). The importance of social identity content in a setting of chronic social conflict: Understanding intergroup relations in Northern Ireland. *British Journal of Social Psychology, 47*(1), 1–21.

Lorenz, E. N. (1963). Deterministic nonperiodic flow. *Journal of the Atmospheric Sciences, 20*, 130–141.

Losada, M. (1999). The complex dynamics of high performance teams. *Mathematical and Computer Modelling, 30*, 179–192.

Losada, M., & Heaphy, E. (2004). The role of positivity and connectivity in the performance of business teams: A nonlinear dynamics model. *The American Behavioral Scientist, 47*, 740–765.

Luhtanen, R., & Crocker, J. (1992). A collective self-esteem scale: Self-evaluation of one's social identity. *Personality and Social Psychology Bulletin, 18*, 302–318.

Luke, A. (1988). *Literacy, textbooks and ideology*. London: Falmer.

Mannix, E. A. (1993). The influence of power, distribution norms, and task meeting structure on resource allocation in small group negotiation. *International Journal of Conflict Management, 4*, 5–23.

Mannix, E. A., Thompson, L. L., & Bazerman, M. H. (1989). Negotiation in small groups. *The Journal of Applied Psychology, 74*, 508–517.

Marshall, M. G., & Gurr, T. R. (2005). *Peace and conflict*. College Park, MD: Center for International Development and Conflict Management.

Marx, K. (1976). *Das Kapital*. Harmondsworth, London: Penguin.

Mason, T. D., Crenshaw, M., McClintock, C., & Walter, B. (2007). *How political violence ends: Paths to conflict de-escalation and termination. APSA task force on political violence and terrorism, Group 3*. Presented at the annual meeting of the American Political Science Association.

Mazur, A., & Booth, A. (1998). Testosterone and dominance in men [target article and commentaries]. *The Behavioral and Brain Sciences, 21*, 353–397.

McClelland, J. L., Rumelhart, D. E., & Hinton, G. E. (1986). Parallel distributed processing: Explorations in the microstructure of cognition. In *Psychological and biological models* (Vol. 2). Cambridge, MA: MIT Press.

McClintock, C. (1976). Social motives in settings of outcome interdependence. In D. Druckman (Ed.), *Negotiations: Social psychological perspective*. Beverly Hills, CA: Sage.

McGrath, J. E., & Kelley, J. R. (1986). *Time and human interaction: Toward a psychology of time*. New York, NY: Guilford.

McNulty, J. K., O'Mara, E. M., & Karney, B. R. (2008). Benevolent cognitions as a strategy of relationship maintenance: "Don't sweat the small stuff"... But it is not all small stuff. *Journal of Personality and Social Psychology, 94*, 631–646.

Mead, G. H. (1934). *Mind, self, and society*. Chicago, IL: University of Chicago Press.

Meinhardt, H. (1995). *The algorithmic beauty in sea shells*. Berlin, Germany: Springer.

Michaels, J. L. (2012). Ph.D. Thesis, Department of Psychology, Florida Atlantic University, Boca Raton, FL.

Milgram, S. (1974). *Obedience to authority: An experimental view*. New York, NY: Harper & Row.

Miller, N. E. (1944). Experimental studies of conflict. In J. M. Hunt (Ed.), *Personality and the behavior disorders* (pp. 431–465). New York, NY: Ronald.

Miller, G. A., Galanter, E., & Pribram, K. (1960). *Plans and the structure of behavior*. New York, NY: Holt.

Miller, K. D., Keller, J. B., & Stryler, M. P. (1989). Ocular dominance columns development: Analysis and simulation. *Science, 245*, 605–615.

Miller, K. E., & Rasmussen, A. (2010). War exposure, daily stressors, and mental health in conflict and post-conflict settings: Bridging the divide between trauma-focused and psychosocial frameworks. *Social Science and Medicine, 70*, 7–16.

Mischel, W., Shoda, Y., & Peake, P. K. (1988). The nature of adolescent competencies predicted by preschool delay of gratification. *Journal of Personality and Social Psychology, 54*, 687–696.

Mischel, W., Shoda, Y., & Rodriguez, M. L. (1989). Delay of gratification in children. *Science, 244*, 933–938.

Monaghan, E., & Glickman, S. (1992). Hormones and aggressiveness behavior. In J. Becker, S. Breedlove, & D. Crews (Eds.), *Behavioral endocrinology* (pp. 261–285). Cambridge, MA: MIT Press.

Mooradian, M., & Druckman, D. (1999). Hurting stalemate or mediation? The conflict over Nagorno-Karabakh, 1990–95. *Journal of Peace Research, 36*(6), 709–727.

Morawski, J. (2000). Social psychology a century ago. *The American Psychologist, 55*(4), 427–430.

Murphy, S. T., & Zajonc, R. B. (1993). Affect, cognition, and awareness: Affective priming with optimal and suboptimal stimulus exposures. *Journal of Personality and Social Psychology, 64*, 723–739.

Musallam, N., Coleman, P. T., & Nowak, A. (2010). Understanding the spread of malignant conflict: A dynamical systems approach. *Peace and Conflict: The Journal of Peace Psychology, 16*, 127–151.

Nan, S. A., Mampilly, Z. C., & Bartoli, A. (Eds.). (2012). *Peacemaking: From practice to theory* (Vol. 2). Santa Barbara, CA: Praeger.

Nets-Zehngut, R. (2008). The Israeli national information center and collective memory of the Israeli–Arab conflict. *The Middle East Journal, 62*(4), 653–670.

Newtson, D., & Nowak, A. (1994). The perception and coupling of behavior waves. In R. R. Vallacher (Ed.), *Dynamical systems in social psychology* (pp. 139–167). San Diego, CA: Academic.

Ng, S. H. (1980). *The social psychology of power*. London: Academic.

Niffenegger, J., & Willer, L. (1998). Friendship behaviors during early childhood and beyond. *Early Childhood Education Journal, 26*(2), 95–99.

Nowak, A. (2004). Dynamical minimalism: Why less is more in psychology. *Personality and Social Psychology Review, 8*, 183–192.

Nowak, A., Bui-Wrzosinska, L., Coleman, P. T., Vallacher, R., Jochemczyk, L., & Bartkowski, W. (2010a). Seeking sustainable solutions: using an attractor simulation platform for teaching multistakeholder negotiation in complex cases. *Negotiation Journal, 26*(1), 49–68.

Nowak, A., Deutsch, M., Bartkowski, W., & Solomon, S. (2010b). From crude law to civil relations: The dynamics and potential resolution of intractable conflict. *Peace and Conflict: Journal of Peace Psychology, 16*(2), 189–209.

Nowak, A., & Lewenstein, M. (1994). Dynamical systems: A tool for social psychology? In R. R. Vallacher & A. Nowak (Eds.), *Dynamical systems in social psychology* (pp. 17–53). San Diego, CA: Academic.

Nowak, A., Lewenstein, M., & Szamrej, J. (1993). Social transitions occur through bubbles. *Scientific American (Polish version), 12*, 16–25.

Nowak A. & Solomon, S. (2006). *Alphabet model of social processes*. Paper presented at the conference "Complexity models of social and economic processes", Torino, Italy.

Nowak, A., Szamrej, J., & Latané, B. (1990). From private attitude to public opinion: A dynamic theory of social impact. *Psychological Review, 97*, 362–376.
Nowak, A., & Vallacher, R. (1998). *Dynamical social psychology*. New York, NY: Guilford Press.
Nowak, A., & Vallacher, R. R. (2001). Societal transition: Toward a dynamical model of social change. In W. Wosinska, R. B. Cialdini, D. W. Barrett, & J. Reykowski (Eds.), *The practice of social influence in multiple cultures* (pp. 151–171). Mahwah, NJ: Lawrence Erlbaum.
Nowak, A., Vallacher, R., Bui-Wrzosinska, L., & Coleman, P. T. (2006). Attracted to conflict: A dynamical perspective on malignant social relations. In A. Golec & K. Skarzynska (Eds.), *Understanding social change: Political psychology in Poland* (pp. 33–49). Haauppague, NY: Nova Science.
Nowak, A., Vallacher, R. R., Kus, M., & Urbaniak, J. (2005). The dynamics of societal transition: Modeling nonlinear change in the Polish economic system. *International Journal of Sociology, 35*, 65–88.
Nowak, A., Vallacher, R. R., Tesser, A., & Borkowski, W. (2000). Society of self: The emergence of collective properties in self-structure. *Psychological Review, 107*, 39–61.
Nowak, A., Vallacher, R. R., & Zochowski, M. (2005). The emergence of personality: Dynamic foundations of individual variation. *Developmental Review, 25*, 351–385.
Operario, D., & Fiske, T. (2001). Ethnic identity moderates perceptions of prejudice: Judgments of personal versus group discrimination and subtle versus blatant bias. *Personality and Social Psychology Bulletin, 27*, 550–561.
Pearce, W. B., & Littlejohn, S. W. (1997). *Moral conflict: When social worlds collide*. Thousand Oaks, CA: Sage.
Perdue, C. W., Dovidio, J. F., Gurtman, M. B., & Tyler, R. B. (1990). Us and them: Social categorization and the process of intergroup bias. *Journal of Personality and Social Psychology, 59*(3), 475.
Peterson, J. B. F., & Flanders, J. L. (2002). Complexity management theory: Motivation for ideological rigidity and social conflict. *Cortex, 38*, 429–458.
Pinker, S. (2011). *The better angels of our nature: Why violence has declined*. New York, NY: Penguin.
Pinkley, R. L., Neale Margaret, A., & Bennett Rebecca, J. (1994). The impact of alternatives to settlement in dyadic negotiation. *Organizational Behavior and Human Decision Processes, 57* (1), 97–116.
Pondy, L. R. (1967). Organizational conflict: Concepts and models. *Administrative Science Quarterly, 12*, 296–320.
Port, R. F., & Van Gelder, T. (1995). *Mind as motion: Explorations in the dynamics of cognition*. Cambridge, MA: MIT Press.
Poston, T., & Stewart, I. (1996). *Catastrophe theory and its applications*. London: Courier Dover.
Powers, W. T. (1973). *Behavior: The control of perception*. Chicago, IL: Aldine.
Praszkier, R., Nowak, A., & Coleman, P. T. (2010). Social entrepreneurs and constructive change: The wisdom of circumventing conflict. *Peace and Conflict, 16*(2), 153–174.
Pruitt, D. G. (1981). *Negotiation behavior*. New York, NY: Academic.
Pruitt, D. G. (1997). Ripeness theory and the Oslo talks. *International Negotiation, 2*(2), 237–250.
Pruitt, D. G. (1998). Social conflict. In D. T. Gilbert, S. T. Fiske, & G. Lindzey (Eds.), *The handbook of social psychology* (4th ed.). New York: McGraw-Hill.
Pruitt, D. G. (2006). *A graphical interpretation of escalation and de-escalation*. Presented at Dynamics and Complexity of Intractable Conflicts, Kazimierz, Poland
Pruitt, D. G., & Kim, S. H. (2004). *Social conflict: Escalation, stalemate, and settlement*. New York, NY: McGraw-Hill.
Pruitt, D. G., & Kimmel, M. J. (1977). Twenty years of experimental gaming: Critique, synthesis, and suggestions for the future. *Annual Review of Psychology, 28*, 363–392.
Pruitt, D. G., & Olczak, P. V. (1995). Resolving seemingly intractable conflict. In B. B. Bunker & J. Z. Rubin (Eds.), *Conflict, cooperation, and justice: Essays inspired by the work of Morton Deutsch*. San Francisco, CA: Jossey-Bass.

Pruitt, D. G., & Rubin, J. Z. (1986). *Social conflict*. New York, NY: McGraw-Hill.
Rahim, M. A. (1983). A measure of styles of handling interpersonal conflict. *The Academy of Management Journal, 26*, 368–376.
Rahim, M. A. (1986). Referent role and styles of handling interpersonal conflict. *The Journal of Social Psychology, 126*, 79–86.
Rawls, J. (1971). *A theory of justice*. Cambridge, CA: Harvard University Press.
Read, S. J., & Miller, L. C. (Eds.). (1998). *Connectionist models of social reasoning and social behavior*. Mahwah, NJ: Lawrence Erlbaum Associates.
Read, S. J., & Miller, L. C. (2002). Virtual personalities: A neural network model of personality. *Personality and Social Psychology Review, 6*, 357–369.
Read, S. J., Vanman, E. J., & Miller, L. C. (1997). Connectionism, parallel constraint satisfaction processes, and Gestalt principles: (Re)introducing cognitive dynamics to social psychology. *Personality and Social Psychology Review, 1*, 26–53.
Reilly, B. (2001). *Democracy in divided societies*. Cambridge, MA: Cambridge University Press.
Reykowski, J. (2006). Deliberative democracy and human nature: An empirical approach. *Political Psychology, 27*(3), 323–346.
Richardson, L. F. (1960). *Arms and insecurity*. Chicago, IL: Quadrangle.
Roccas, S., & Brewer, M. B. (2002). Social identity complexity. *Personality and Social Psychology Review, 6*, 88–106.
Rosenblatt, A., Greenberg, J., Solomon, S., Pyszczynski, T., & Lyon, D. (1989). Evidence for terror management theory I: The effects of mortality salience on reactions to those who violate or uphold cultural values. *Journal of Personality and Social Psychology, 57*, 681–690.
Rouhana, N. N., & Bar-Tal, D. (1998). Psychological dynamics of intractable ethnonational conflicts: The Israeli–Palestinian case. *The American Psychologist, 53*, 761–770.
Rubin, J. Z., & Levinger, G. (1995). Levels of analysis: In search of generalizable knowledge. In B. B. Bunker & J. Z. Rubin (Eds.), *Cooperation, conflict, and justice: Essays inspired by the work of Morton Deutsch* (pp. 13–38). New York, NY: Sage.
Rubin, J. Z., Pruitt, D. G., & Kim, S. H. (1994). *Social conflict: Escalation, stalemate, and settlement* (2nd ed.). New York, NY: McGraw-Hill.
Ruelle, D. (1989). *Elements of differentiable dynamics and bifurcation theory*. New York, NY: Academic.
Saltzman, B. (1962). Finite amplitude free convection as an initial value problem—I. *Journal of the Atmospheric Sciences, 19*, 329–341.
Sandole, D. J. D. (1999). *Capturing the complexity of conflict: Dealing with violent ethnic conflicts in the post-cold war era*. New York, NY: Pinter.
Satish, U. (1997). Behavioral complexity: A review. *Journal of Applied Social Psychology, 27*, 2047–2067.
Saunders, A. D. (1997). *Channel defenses*. London: BT Batsford Ltd.
Savelsberg, J., & King, R. D. (2005). Institutionalizing collective memories of hate: Law and law enforcement in Germany and the United States. *The American Journal of Sociology, 111*(2), 579–616.
Schelling, T. C. (1960). *The strategy of conflict*. Cambridge, MA: Harvard University Press.
Schulz-Hardt, S., Mojzisch, A., & Vogelgesang, F. (2008). Dissent as a facilitator: Individual- and group-level effects on creativity and performance. In C. K. W. De Dreu & M. J. Gelfand (Eds.), *The psychology of conflict and conflict management in organizations* (pp. 149–177). New York, NY: Taylor and Francis Group/Lawrence Erlbaum Associates.
Schuster, H. G. (1984). *Deterministic chaos*. Vienna, Austria: Physik.
Schwartz, I. B. (1992). Small amplitude, long period outbreaks in seasonally driven epidemics. *Journal of Mathematical Biology, 30*, 473–491.
Seager, T. (2008). The sustainability spectrum and the sciences of sustainability. *Business Strategy and the Environment, 17*, 444–453.
Seligman, M. E. P., Steen, T. A., & Park, N. (2005). Positive psychology progress: Empirical validation of interventions. *The American Psychologist, 60*, 410–421.

Shamir, M., & Sagiv-Schifter, T. (2006). Conflict, identity, and tolerance: Israel in the Al-Aqsa Intifada. *Political Psychology, 27*, 569–595.

Sherif, M., Harvey, O. J., White, J., Hood, W., & Sherif, C. W. (1961). *Intergroup conflict and cooperation: The robber's cave experiment*. Norman, OK: Institute of Intergroup Relations, University of Oklahoma.

Shnerb, N. M., Louzoun, Y., Bettelheim, E., & Solomon, S. (2000). The Importance of being discrete: life always wins on the surface. *Proceedings of the National Academy of Science, 97*, 10322.

Shoda, Y., Tiernan, L., & Mischel, W. (2002). Personality as a dynamical system: Emergence of stability and distinctiveness from intra- and interpersonal interactions. *Personality and Social Psychology Review, 6*, 316–325.

Sidanius, J., Levin, S., Federico, C., & Pratto, F. (2001). Legitimizing ideologies: The social dominance approach. In J. Jost & B. Major (Eds.), *The psychology of legitimacy: Emerging perspectives on ideology, justice, and intergroup relations* (pp. 307–331). Cambridge, MA: Cambridge University Press.

Sidanius, J., Pratto, F., van Laar, C., & Levin, S. (2004). Social dominance theory: Its agenda and method. *Political Psychology, 25*, 845–880.

Smith, E. R. (1996). What do connectionism and social psychology offer each other? *Journal of Personality and Social Psychology, 70*, 893–912.

Smith, L. B., & Thelen, E. (1993). *A dynamic systems approach to development: Applications*. Cambridge, MA: MIT Press.

Staub, E. (2001). Genocide and mass killing: Their roots and prevention. In D. J. Christie, R. V. Wagner, & D. D. Winter (Eds.), *Peace, conflict and violence: Peace psychology for the 21st century* (pp. 76–86). Upper Saddle River, NJ: Prentice Hall.

Stedman, S. J., Rothchild, D., & Cousens, E. M. (2002). *Ending civil wars: The implementation of peace agreements*. Boulder, CO: Lynne Rienner.

Sternberg, R. J., & Dobson, D. M. (1987). Resolving interpersonal conflicts: An analysis of stylistic consistency. *Journal of Personality and Social Psychology, 52*, 794–812.

Strogatz, S. (1994). *Nonlinear dynamics and Chaos*. Cambridge, MA: Perseus Books.

Strogatz, S. (2003). *Sync: The emerging science of spontaneous order*. New York, NY: Hyperion Books.

Suhrke, A., & Samset, I. (2007). What's in a figure? Estimating recurrence of civil war. *International Peacekeeping, 14*(2), 195–203.

Sullivan, J. L., Piereson, J., & Marcus, G. E. (1993). *Political tolerance and American democracy*. Chicago, IL: University of Chicago Press.

Sullivan, J. L., Shamir, M., Walsh, P., & Roberts, N. S. (1985). *Political tolerance in context: Support for un-popular minorities in Israel New Zealand, and the United States*. Boulder, CO: Westview.

Swann, W. B., Jr. (1990). To be adored or to be known? The interplay of self-enhancement and self-verification. In E. T. Higgins & R. M. Sorrentino (Eds.), *Handbook of motivation and cognition: Foundations of social behavior* (pp. 408–448). New York, NY: Guilford.

Swann, W. B., Jr., Hixon, J. G., Stein-Seroussi, A., & Gilbert, D. (1990). The fleeting gleam of praise: Cognitive processes underlying behavioral reactions to self-relevant feedback. *Journal of Personality and Social Psychology, 59*, 17–26.

Tajfel, H. (1970). Experiments in intergroup discrimination. *Scientific American, 223*, 96–102.

Tajfel, H. (1974). Social identity and intergroup behaviour. *Social Science Information, 13*, 65–93.

Tajfel, H. (1979). Individuals and groups in social psychology. *The British Journal of Social and Clinical Psychology, 18*, 183–190.

Tajfel, H. (1982). *Social identity and intergroup relations*. Cambridge, England: Cambridge University Press.

Tajfel, H., & Turner, J. C. (1979). An integrative theory of intergroup conflict. In W. G. Austin & S. Worchel (Eds.), *The social psychology of intergroup relations* (pp. 33–47). Monterey, CA: Brooks/Cole.

Tesser, A. (1978). Self-generated attitude change. In L. Berkowitz (Ed.), *Advances in experimental social psychology* (Vol. 11, pp. 85–117). New York: Academic.

Tesser, A., & Achee, J. (1994). Aggression, love, conformity and other social psychological catastrophes. In R. Vallacher & A. Nowak (Eds.), *Dynamical systems in social psychology* (pp. 95–109). San Diego, CA: Academic.

Tesser, A., Martin, L., & Cornell, D. (1996). On the substitutability of self-protective mechanisms. In P. M. Gollwitzer & J. A. Bargh (Eds.), *The psychology of action* (pp. 48–68). New York, NY: Guilford.

Thelen, E., & Smith, L. B. (Eds.). (1994). *A dynamic systems approach to the development of cognition and action*. Cambridge, MA: MIT Press/Bradford Books.

Thibaut, J. W., & Kelley, H. H. (1959). *The social psychology of groups*. New York, NY: Wiley.

Thom, R. (1975). *Structural stability and morphogenesis*. New York, NY: Addison-Wesley.

Thomas, K. W. (1976). *Thomas-Kilmann conflict model*. Tuxedo, NY: Xicom/Consulting Psychologists Press.

Thorndike, E. L. (1920). A constant error in psychological ratings. *The Journal of Applied Psychology, 4*, 25–29.

Tice, D. M., Baumeister, R. F., Shmueli, D., & Muraven, M. (2007). Restoring the self: Positive affect helps improve self-regulation following ego depletion. *Journal of Experimental Social Psychology, 43*, 379–384.

Tice, D. M., Bratslavsky, E., & Baumeister, R. F. (2001). Emotional distress regulation takes precedence over impulse control: If you feel bad, do it! *Journal of Personality and Social Psychology, 80*, 53–67.

Tickle-Degnen, L., & Rosenthal, R. (1987). Group rapport and nonverbal behavior. *Review of Personality and Social Psychology, 9*, 113–136.

Tjosvold, D. (1991). *The conflict positive organization*. Reading, MA: Addison-Wesley.

Tjosvold, D., & Johnson, D. W. (1983). *Productive conflict management: Perspectives for organizations*. New York, NY: Irvington.

Triandis, H. C. (1995). *Individualism and collectivism*. Boulder, CO: Westview Press.

Turcotte, D. L., & Rundle, J. B. (2002). Self-organized complexity in the physical, biological, and social sciences. *Proceedings of the National Academy of Sciences of the United States of America, 99*(Suppl 1), 2463–2465.

Turner, J. C., Reynolds, K. J., & Reynolds, K. J. (2001). The social identity perspective in intergroup relations: Theories, themes and controversies. In R. J. Brown (Ed.), *Handbook of social psychology* (Intergroup processes, Vol. 4, pp. 133–152). Oxford, England: Blackwell.

Turvey, M. T. (1990). Coordination. *The American Psychologist, 4*, 938–953.

UNFPA. (2006). *UNFPA state of the world population. United Nations population fund*. New York, NY: United Nations, Department of Public Information.

United Nations. (2004). *A more secure world: our shared responsibility. Report of the United Nations high-level panel on threats, challenges and change*. New York, NY: United Nations.

United Nations Assistance Mission for Iraq (UNAMI). (2006). *Human rights report 1 May–30 June 2006*. New York, NY: UNAMI.

Ury, W. (1991). *Getting past no*. New York, NY: Bantam Books.

Vallacher, R. R., Coleman, P. T., Nowak, A., & Bui-Wrosinska, L. (2010). Rethinking intractable conflict: The perspective of dynamical systems. *The American Psychologist, 65*, 262–278.

Vallacher, R. R., & Nowak, A. (Eds.). (1994a). *Dynamical systems in social psychology*. San Diego, CA: Academic.

Vallacher, R. R., & Nowak, A. (1994b). The stream of social judgment. In R. R. Vallacher & A. Nowak (Eds.), *Dynamical systems in social psychology* (pp. 251–275). San Diego, CA: Academic.

Vallacher, R. R., & Nowak, A. (1997). The emergence of dynamical social psychology. *Psychological Inquiry, 4*, 73–99.

Vallacher, R. R., & Nowak, A. (1999). The dynamics of self-regulation. In R. S. Wyer Jr. (Ed.), *Advances in social cognition* (Vol. 12, pp. 241–259). Mahwah, NJ: Lawrence Erlbaum.

Vallacher, R. R., & Nowak, A. (2006). Coherence in human experience and psychological science. In P. A. M. Van Lange (Ed.), *Bridging social psychology: Benefits of transdisciplinary approaches* (pp. 77–82). Mahwah, NJ: Lawrence Erlbaum Associates.
Vallacher, R. R., & Nowak, A. (2007). Dynamical social psychology: Finding order in the flow of human experience. In A. W. Kruglanski & E. T. Higgins (Eds.), *Social psychology: Handbook of basic principles* (pp. 734–758). New York, NY: Guilford.
Vallacher, R. R., Nowak, A., & Kaufman, J. (1994). Intrinsic dynamics of social judgment. *Journal of Personality and Social Psychology, 67*, 20–34.
Vallacher, R. R., Nowak, A., & Zochowski, M. (2007). Dynamics of social coordination: The synchronization of internal states in close relationships. In P. Hauf & F. Forsterling (Eds.), *Making minds: The shaping of human minds through social context* (pp. 31–46). Amsterdam: John Benjamins.
Vallacher, R. R., Read, S. J., & Nowak, A. (Eds.). (2002). *The dynamical perspective in social psychology* (Personality and Social Psychology Review, Vol. 6, pp. 264–388).
Vallacher, R. R., & Selz, K. (1991). Who's to blame? Action identification in allocating responsibility for alleged rape. *Social Cognition, 9*, 194–219.
Vallacher, R. R., & Wegner, D. M. (1987). What do people think they're doing? Action identification and human behavior. *Psychological Review, 94*, 3–15.
Vallacher, R. R., & Wegner, D. M. (2012). Action identification theory. In P. A. M. Van Lange, A. W. Kruglanski, & E. T. Higgins (Eds.), *Handbook of theories of social psychology* (pp. 327–348). Thousand Oaks, CA: Sage.
Van de Vliert, E., & Kabanoff, B. (1990). Toward theory-based measures of conflict management. *The Academy of Management Journal, 33*, 199–209.
van Geert, P. (1991). A dynamic systems model of cognitive and language growth. *Psychological Review, 98*, 3–53.
Van Lange, P. A. M., De Cremer, D., Van Dijk, E., & Van Vugt, M. (2007). Self-interest and beyond: Basic principles of social interaction. In A. W. Kruglanski & E. T. Higgins (Eds.), *Social psychology: Handbook of basic principles* (2nd ed.). New York, NY: Guilford Press.
Varshnay, W. (2002). *Ethnic conflict and civic life*. New Haven, CT: Yale University Press.
Vohs, K. D., Baumeister, R. F., & Ciarocco, N. (2005). Self-regulation and self-presentation: Regulatory resource depletion impairs impression management and effortful self-presentation depletes regulatory resources. *Journal of Personality and Social Psychology, 88*, 632–657.
Vohs, K. D., Baumeister, R. F., Schmeichel, B. J., Twenge, J. M., Nelson, N. M., & Tice, D. M. (2008). Making choices impairs subsequent self-control: A limited resource account of decision-making, self-regulation, and active initiative. *Journal of Personality and Social Psychology, 94*, 883–898.
Vohs, K. D., & Heatherton, T. F. (2000). Self-regulatory failure: A resource-depletion approach. *Psychological Science, 11*, 249–254.
Von Bertalannfy, L. (1968). *General system theory*. New York, NY: Braziller.
Von Neumann, J., & Morgenstern, O. (1944). *Theory of games and economic behavior*. Princeton, NJ: Princeton University Press.
Voronov, M., & Coleman, P. T. (2003). Beyond the ivory towers: Organizational power practices and a "practical" critical postmodernism. *The Journal of Applied Behavioral Science, 39*(2), 169–185.
Waldrop, M. M. (1992). *Complexity: The emerging science at the edge of order and chaos*. New York, NY: Schuster.
Webster, D. M., & Kruglanski, A. W. (1994). Individual differences in need for cognitive closure. *Journal of Personality and Social Psychology, 67*, 1049–1062.
Wegner, D. M. (1989). *White bears and other unwanted thoughts: Suppression, obsession, and the psychology of mental control*. New York, NY: Viking.
Wegner, D. M. (1994). Ironic processes of mental control. *Psychological Review, 101*, 34–52.
Wegner, D. M., Erber, R., & Zanakos, S. (1993). Ironic processes in the mental control of mood and mood-related thought. *Journal of Personality and Social Psychology, 65*, 1093–1104.

Wegner, D. M., & Pennebaker, J. W. (Eds.). (1993). *Handbook of mental control*. Englewood Cliffs, NJ: Prentice-Hall.

Wegner, D. M., Schneider, D. J., Carter, S. R., & White, T. L. (1987). Paradoxical effects of thought suppression. *Journal of Personality and Social Psychology, 53*, 5–13.

Wegner, D. M., & Vallacher, R. R. (1977). *Implicit psychology: An introduction to social cognition*. New York, NY: Oxford University Press.

Wegner, D. M., Vallacher, R. R., Macomber, G., Wood, R., & Arps, K. (1984). The emergence of action. *Journal of Personality and Social Psychology, 46*, 269–279.

Weiner, N. (1948). *Cybernetics, or control and communication in the animal and the machine*. Cambridge, MA: MIT Press.

Weingart, L. R., Bennet, R. J., & Brett, J. M. (1993). The impact of consideration of issues and motivational orientation on group negotiation process and outcome. *The Journal of Applied Psychology, 78*, 504–517.

Weisbuch, G. (1992). *Complex system dynamics*. Redwood City, CA: Addison-Wesley.

Wessells, M. G. (2007). Post-conflict healing and reconstructing for peace: The power of social mobilization. In J. D. White & A. J. Marsella (Eds.), *Fear of persecution: Global human rights, international law and human well being* (pp. 258–278). Lanham, MD: Lexington Books.

Wiener, N. (1948). *Cybernetics: Control and communication in the animal and the machine*. Cambridge, MA: MIT Press.

Wilson, T. D. (2002). *Strangers to ourselves: Discovering the adaptive unconscious*. Cambridge, MA: Belknap Press of Harvard University Press.

Winter, D. G. (2007). The role of motivation, responsibility, and integrative complexity in crisis escalation: Comparative studies of war and peace crises. *Journal of Personality and Social Psychology, 92*, 920–937.

Wolfram, S. (2002). *A new kind of science*. Champaign, IL: Wolfram Media.

Worchel, S. (1986). The role of cooperation in reducing intergroup conflict. In S. Worchel & W. Austin (Eds.), *The psychology of intergroup relations* (pp. 288–304). Chicago, IL: Nelson-Hall.

Zartman, I. W. (1986). Ripening conflict, ripe moment, formula and mediation. In D. BenDahmane & J. McDonald (Eds.), *Perspectives on negotiation*. Washington, DC: GPO.

Zartman, I. W. (1989). *Ripe for resolution: Conflict and intervention in Africa* (2nd ed.). New York, NY: Oxford University Press.

Zartman, I. W. (2000). Ripeness: The hurting stalemate and beyond. In P. Stern & D. Druckman (Eds.), *International conflict resolution after the cold war* (pp. 225–250). Washington, DC: National Academy Press.

Zartman, I. W. (2003). Negotiating with terrorists. *International Negotiation, 8*, 443–450.

Zartman, W. A. (2005). Need, creed, and greed in intrastate conflict. In C. J. Arnson, W. A. Zartman, W. A. Zartman, & W. A. Zartman (Eds.), *Rethinking the economics of war: The intersection of need, creed, and greed*. Baltimore, MD: Johns Hopkins University Press.

Zartman, I. W., & Faure, G. O. (2005). *Escalation and negotiation in international conflicts* (1st ed.). New York, NY: Cambridge University Press.

Zimbardo, P. G. (1970). The human choice: Individuation, reason, and order versus deindividuation, impulse, and chaos. In W. J. Arnold & D. Levine (Eds.), *Nebraska symposium on motivation, 1969* (pp. 237–307). Lincoln: University of Nebraska Press.

Zochowski, M., & Liebovitch, L. S. (1997). Synchronization of the trajectory as a way to control the dynamics of the coupled system. *Physical Review E, 56*, 3701.

Zochowski, M., & Liebovitch, L. S. (1999). Self-organizing dynamics of coupled map systems. *Physical Review E, 59*, 283.

Index

R.R. Vallacher et al., *Attracted to Conflict: Dynamic Foundations of Destructive Social Relations*, Peace Psychology Book Series, DOI 10.1007/978-3-642-35280-5,

D